Advances in Meat Processing Technologies: Modern Approaches to Meet Consumer Demand

Authored by

Daneysa L. Kalschne

Marinês P. Corso

&

Cristiane Canan
Universidade Tecnológica Federal do Paraná
Department of Food, Medianeira, Paraná
Brazil

Advances in Meat Processing Technologies:
Modern Approaches to Meet Consumer Demand

Authors: Daneysa L. Kalschne, Marinês P. Corso and Cristiane Canan

ISBN (Online): 978-981-14-7019-6

ISBN (Print): 978-981-14-7017-2

ISBN (Paperback): 978-981-14-7018-9

need for a court order if at any point you breach any terms of this License Agreement. In no event will any delay or failure by Bentham Science Publishers in enforcing your compliance with this License Agreement constitute a waiver of any of its rights.

3. You acknowledge that you have read this License Agreement, and agree to be bound by its terms and conditions. To the extent that any other terms and conditions presented on any website of Bentham Science Publishers conflict with, or are inconsistent with, the terms and conditions set out in this License Agreement, you acknowledge that the terms and conditions set out in this License Agreement shall prevail.

Bentham Science Publishers Pte. Ltd.
80 Robinson Road #02-00
Singapore 068898
Singapore
Email: subscriptions@benthamscience.net

BENTHAM SCIENCE

CONTENTS

FOREWORD

The aim of this book is to provide an approach to recent improvements in conventional meat products' processing and the development and/or use of new technologies to students, researchers, and lecturers in the meat science and technology field, as well as professionals in the meat industry. Needs and demands by consumers for meat-based products have guided meat processing plants to adapt themselves to this new reality by implementing innovative technologies and ensuring that meat products are stable and microbiologically safe, nutritionally healthy, with good sensory characteristics, practical and of quick preparation, with compatible and affordable costs. Beef, pork, and poultry are the ones most used in meat products preparation. The use of additives and traditional techniques, such as fermentation and drying, salting, smoking, and cooking, contribute to the production of safe meat products. Chapter 1 provides an overview regarding the importance of the meat-based market in the world, as well as overall meat processing aspects. Chapters 2 to 7 describe conventional and emerging technologies presenting relevant and promising data applied in the processing of dry-fermented meat products, emulsified, salted and non-emulsified thick mass, marinated, and restructured products, and advancements in fresh meat tenderness. These chapters discuss the QDS (quick-dry-slice) technology that aims at reducing the processing time of sliced matured and/or fermented meat products; the use of ultrasound or high pressure to reduce salt contents of meat products as well as improve emulsion stability and replace pork fat; and new cooking methods such as microwaving, radio frequency and ohmic heating, the latter two could be used for reducing cooking time and production costs. Also, technologies such as pulsed light, ultraviolet light, irradiation, ultrasound, and high hydrostatic pressure may reduce microbial contamination of meat and meat-based products. Furthermore, the application of ultrasound waves could augment fresh meat tenderness. The use of such new technologies would reduce additives content, posing economic benefits such as reduced preparation time and reduced energy and water consumption. In addition, safety and quality are secured and the processed products have sensory characteristics similar to products obtained using traditional technologies. A specific chapter describes strategies for the development and production of meat analogues with nutritional and sensory properties typical of meat products to meet the demands of vegetarians and vegans alike. New approaches to the use of alternative ingredients in order to obtain meat-based products with technological, microbiological, and sensory quality are discussed in the last chapter. These new ingredients aim at replacing animal fat, reducing sodium chloride, and chemical additives, such as nitrites and synthetic antioxidants. The advancements in processing technologies and/or development of new meat-based products covered in this book were based on relevant literature and experiences of the authors with nearly two decades of experience in lecturing and researching, as well as technical-scientific interaction with meat processing plants. This book on Advances in Meat Processing Technologies ought to provide the readers with further knowledge regarding recent developments in the new technology for the full meat processing chain.

Elza Iouko Ida
Centro de Ciências Agrárias, Universidade Estadual de Londrina, Rodovia Celso Garcia
Cid, Postal code 86057-970, P.O. Box 10.001, Londrina, Paraná
Brazil

PREFACE

This ebook titled 'Cutting Edge Techniques in Biophysics, Biochemistry and Cell Biology: From Principle to Applications' provides principles of basic and analytical techniques in biophysics, biochemistry and cell biology and their potential applications in biomedical research. The volume covers the basics of *in-vitro* cell culture techniques, such as flow cytometry including FACS, Real time PCR based disease diagnosis as well as gene expression analysis, X ray crystallography, RNA sequencing and their various biomedical applications, including drug screening, disease model, functional assays, disease diagnosis, gene expression analysis, and protein structure determination. This is a valuable resource for biomedical students, cellular and molecular biologists who want to be an eminent biomedical researcher.

We strongly believe that this book is a reader's delight providing a comprehensive understanding on cell biology, macromolecular structure, *in silico* studies and functioning. It is an excellent introduction for students towards cell biology and biophysics that clearly help them to develop a wider perspective of the field. The idea to encompass knowledge covering from what is known to what is unknown in the writings from the experts in the field results in a concerted effort to justify the various topics. We sincerely hope our efforts will be embraced by students with appreciation and enthusiasm for learning.

CONSENT FOR PUBLICATION

Not applicable.

CONFLICT OF INTEREST

The authors declare that they have no conflict of interest.

ACKNOWLEDGEMENTS

The authors would like to acknowledge the following Brazilian finance agencies; Conselho Nacional de Desenvolvimento Científico e Tecnológico (CNPq), Coordenação de Aperfeiçoamento de Pessoal de Nível Superior (CAPES), Fundação Araucária, Universidade Tecnológica Federal do Paraná (UTFPR), the Post-Graduate Program in Food Technology (PPGTA), and Universidade Estadual de Londrina and the Post-Graduate Program in Food Science for their support in our research and for elaboration of this book. The authors would also like to acknowledge the meat processing plants located in Western Paraná State for the collaboration and partnerships in recent years, the under-graduate and post-graduate students for their motivation, dedication, companionship, and fun, and Dr. Paulo Stival Bittencourt for his support, elaborating the figures section. Last but not the least, we would like to thank two brilliant Brazilian researchers, Dr. Nelcindo Nascimento Terra from the Meat Technology field and Dr. Massami Shimokomaki from the Meat Science field (both in memoriam), for their great examples that have encouraged and inspired us throughout our careers and for the elaboration of this book.

Daneysa L. Kalschne
Marinês P. Corso
&
Cristiane Canan
Universidade Tecnológica Federal do Paraná
Department of Food, Medianeira, Paraná
Brazil

The Meat Industrialization Scenario: Production, Consumption, and Product Changes Driven by Consumers

Abstract: Meat is consumed worldwide and from ancient times to the present day, the development of meat products and processes has been driven by meat consumers. Beef, pork and poultry are the most common varieties of meat preferred for human consumption and the emergence of meat processing industrialization has allowed the use of the entire carcass, the development of different products, and microbiological safety due to advances in the technology used. Meat consumers have been driving production trends for meat products that aim at acceptable nutritional, physico-chemical, microbiological and sensory characteristics of the final product. In this context, the development of emerging technologies for meat industrialization aims at improving conventional processes, offering new and suitable solutions to the meat industry to meet consumer demand.

Keywords: Additive, Emergent Technologies, Healthier Products, High Hydrostatic Pressure, Irradiation, Microwave, Meat Analogues, Microbiology Safety, Natural Ingredients, Ohmic Heating, Pulsed Light, Radio-Frequency, Ultrasound.

INTRODUCTION

Meat is consumed worldwide. In the past, it was associated with strength and power, a typical illustration of one's social status. Nowadays, this perception is still valid and such a context has persisted over the years [1]. Meat sensory properties, such as color, aroma, texture, flavor, juiciness, and chewability [2] have attracted ever more demanding consumers.

Furthermore, meat is recognized by its nutritional properties, consisting of 70% water, which is affected by the lipid content ranging from 8% to 20%, 18% to 23% protein, 1% carbohydrates, and 1 to 1.2% minerals [3, 4]. The protein content has been highlighted due to amino acids' high biological value especially those essential to the human diet [4 - 6].

The content and composition of meat lipids vary according to aspects such as spe-

Daneysa L. Kalschne, Marinês P. Corso & Cristiane Canan

cies, breed, gender, age, feed (especially for pork and chicken), nutritional status, as well as the animal's activity level [7 - 10]. Overall, meat has high saturated fatty acid levels to keep the fat tissue healthy, while phospholipids

possess higher unsaturated fatty acids levels conferring greater fluidity to the muscular cells' plasmatic membrane [3].The small portion (<1%) of carbohydrates is represented mainly by glycogen that is used as a muscle energy source [3, 4].

Meat is considered an excellent source of water-soluble vitamins such as thiamine, riboflavin, niacin, vitamin B_6, and vitamin B_{12}. For the mineral content, red meat is famous due to its iron content combined with myoglobin content. In addition, potassium, phosphorus and magnesium are also abundant in meat [3, 4, 11]. In addition to the importance of calcium in muscle contraction, considerable amounts of such minerals are only found in mechanically processed meat due to the presence of microscopic bone fragments [3].

In 2018, worldwide meat production increased by 1.0% to 327 Mt, reflecting an increase in bovine (hereafter "beef"), pork and poultry meat production, with rather modest gains for lamb. The world's meat exports (excluding live animals and processed products) are forecasted to increase 18% more by 2028 than in the base period. This represents a decline in the 1.4% average annual meat trading growth rate compared to 3% during the previous decade. However, the total meat output traded share on the global market could increase slightly [12].

According to the data from the United States Department of Agriculture (USDA) [13], worldwide beef production for 2020 could reach 61,861 Mt (carcass weight equivalent), led by the USA and followed by Brazil, the European Union and China. For pork, the forecast for 2020 indicates a 95,223 Mt (carcass weight equivalent) production, led by China and followed by the European Union, USA, and Brazil. For chicken, the production in 2020 could reach 103,498 Mt (ready-to-cook equivalent), led by the USA and followed by China, Brazil and the European Union.

Fig. (1) presents the worldwide per capita meat consumption considering beef, pork, and poultry meat per continent. The data were provided by the Organization for Economic Co-operation and Development (OECD) [14]. It was observed that the leading meat producers (including beef, pork and chicken) are also the leading consumers.

The meat marketing success depends on continuous innovation and increased production of high-quality products. It is important to point out that meat consumption patterns are constantly changing. These changes are associated not only with wide-ranging, socio-economic, and cultural trends, but also with specific lifestyles of increasingly diversified consumer groups and the use of natural resources.

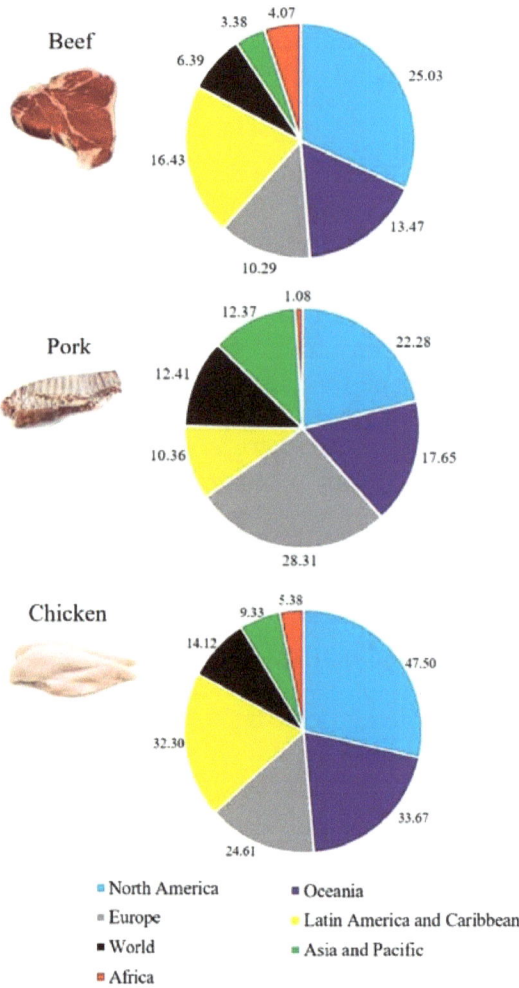

Fig. (1). Meat per capita consumption worldwide (%) (data from OECD [14]).

In this context, good taste, convenience orientation, and meat avoidance are some important meat consumption trends that put pressure on policymakers aiming to influence consumer behavior and achieve more sustainable consumption [15]. It should be emphasized that meat products' sales are subdivided into fresh and processed ones and the decision to purchase either one or the other is affected by the consumer profile. Consumer preferences differ according to the type of meat (beef, pork, or chicken). In Latin America, for instance, beef is mostly sold in its fresh form as opposed to pork that is mainly sold in its processed form (bacon, sausage, mortadella *etc*); the only fresh pork cuts sold on a large scale are chops, ribs and leg [16]. Chicken meat has a greater balance in terms of fresh and processed products' sales; however, its sale in the fresh form prevails. A survey in Europe showed that processed meat products' consumption was around 54% by men and 24% by women, reinforcing that many factors influence meat preferences [17]. Another aspect of meat consumption trends is that higher income allows for a greater meat purchasing power, which is the case in both the more developed and developing countries; meat consumption in developing countries increases as purchasing power increases [18].

Consumers always look for convenient food products with new/exciting flavors, textures, *etc*. It is not enough for the consumer that the food only meets the nutritional aspects; consumers also look for a variety when eating food [19]. Moreover, the failure rate of innovative products launched in the market is high, reinforcing the consumer decision-making power in the search for more suitable products that meet their expectations. The consumers' choice for processed meat products represents a substantial meat market share, and it deserves further study [20].

Overall, processed meat products refer to meat transformed by conventional processes such as salting, curing, fermentation, cooking, smoking, among others allowing for achieving the sensory characteristics of products. Nowadays, there are a wide range of meat products varying in terms of the raw materials and ingredients used, production processes as well as chemical compositions. All in all, they are an important source of protein, B-complex vitamins and hematinic iron, among other nutrients.

Processed meat products could be grouped into categories as per the similarity of processes to which they are subjected. The most widespread categories worldwide include dry-fermented products (*i.e.* raw hams, salami, *etc*.), salted products (jerky, jerked beef, bacon, *etc*.), emulsified sausages (frankfurters and mortadellas), non-emulsified sausages (fresh and cooked thick dough), marinated and restructured (seasoned, nuggets, *etc*.) and more recently meat analogues that are innovative products processed using techniques transforming non-meat protein and other compounds into a fibrous material with meat-like characteristics.

At first, pork was the main processed meat; however, over the past four decades, the growth in poultry production has resulted in several meat products. The mild taste of chicken contributes to the development of practical and easy to prepare products (marinated/seasoned/breaded products). Such an alternative stimulates the inclusion of chicken in everyday meals, pleasing consumers. Also, traditional products such as frankfurters, mortadella, burgers, among others, could also be chicken-meat-made.

Historically, lower product prices have contributed to making poultry and pork meat the favorite consumers' choices in developing countries [16]. On the other hand, increased income levels allow consumers to gradually diversify their meat consumption into more expensive meat varieties such as beef and lamb. Other factors have also influenced meat consumption trends, such as religious beliefs, cultural norms, urbanization, as well as environmental, ethical, and health concerns [1]. Consumer demands for healthier foodstuffs have increased over the last few years and the food industry has been up to the challenge [21].

Alternatives to industrialize meat have been developed to ensure nutritional, sensory and safety aspects and emerging technologies are playing an important role in food production, improving existing traditional technologies. These so-called green technologies use either physical, chemical, or biological methods more efficiently with lower electric power, chemical, and water consumption as well as lower wastewater production. These incorporate drying time reduction during the QDS (Quick-Dry-Slice) process used for mature and/or fermented sliced meat products, salt content reduction in meat products using ultrasound or high pressure, new cooking methods such as microwave, radiofrequency and ohmic heating, reducing processing time and cost. Also, technologies such as pulsed light, ultraviolet light, irradiation, ultrasound and high hydrostatic pressure have been used to reduce microbial contamination in meat and meat products.

Moreover, efforts have been made to replace animal fat and chemical additives/ingredients (nitrites, salt, synthetic antioxidants, phosphates and others) linked to health problems. Significant innovations have been achieved and made available to the meat industry to ensure meat products with microbiologically safe, stable, and sensorially attractive characteristics, similar to those obtained by conventional processes at reasonable costs. Such advances have provided a healthier approach to meet fierce consumer demands of meat products that have been the driving force on developing innovative technological processes.

Today, the trend for vegetarian or vegan food is notable. In this context, studies have sought to develop animal-ingredient-free products but with the appearance and sensory characteristics similar to meat products. For developing these new products, important innovations have already been achieved to meet consumers' expectations. Consumers have changed their perceptions of animal food production and consumption and a great number have moved to either vegetarian or vegan diets. Thus, the development of meat analogues has represented a significant approach in the past 5 years. Such consumers have already been attracted to consuming meat, but have decided to opt for an alternative diet. Therefore, they look for the same sensory properties and functionality in meat analogues that meat provides, opening a new market opportunity.

The subsequent chapters describe the use, advantages and disadvantages of most studied emerging technologies for meat products, and the advances in additives use and the meat analogues approach.

FINAL CONSIDERATIONS

Meat is a well-established product in the human diet worldwide and consumers have driven the production and processing development of meat and meat products traded. Emerging technologies could be used either alone or combined with existing conventional technologies to meet consumers' demands ensuring the food's nutritional, physico-chemical, microbiological and sensory characteristics. Food scientists have the role of meeting the expectations of the consumers and the industry, influencing production, industrialization, and marketing of food by offering suitable solutions.

REFERENCES

[1]   E. Y. Chan, and N. Zlatevska, "Jerkies, tacos, and burgers: Subjective socioeconomic status and meat preference", *Appetite.,* vol. 132, no. 1, pp. 257-266, 2019.
[http://dx.doi.org/10.1016/j.appet.2018.08.027] [PMID: 30172366]

[2]   V. Joshi, and S. Kumar, "Meat Analogues: Plant based alternatives to meat products- A review", *Int. J. Food Ferment. Technol.,* vol. 5, no. 2, p. 107, 2015.
[http://dx.doi.org/10.5958/2277-9396.2016.00001.5]

[3]   G. Strasburg, Y. Xiong, and W. Chiang, "Fisiologia e Química dos tecidos musculares comestíveis", In: *Química de Alimentos de Fennema,* S. Damodaran, K. L. Parkin, O. R. Fennema, Eds., Artmed: Porto Alegre, 2010, pp. 719-757.

[4]   R. S Ahmad, A. Imran, and M. B. Hussain, "Nutritional Composition of Meat", In: *Meat Science and Nutrition,* M. S Arshad, Ed., IntechOpen: London, 2018, pp. 61-74.
[http://dx.doi.org/10.5772/intechopen.77045]

[5]   D. Magnoni, and I. Pimentel, "A importância da carne suína na nutrição humana", Available: http://www.abcs.org.br/attachments/0994.pdf. [Accessed Jun. 10, 2018].

[6]   G. Wu, H.R. Cross, K.B. Gehring, J.W. Savell, A.N. Arnold, and S.H. McNeill, "Composition of free and peptide-bound amino acids in beef chuck, loin, and round cuts", *J. Anim. Sci.,* vol. 94, no. 6, pp. 2603-2613, 2016.
[http://dx.doi.org/10.2527/jas.2016-0478] [PMID: 27285936]

[7]   D.T. Utama, Y.J. Kim, H.S. Jeong, J. Kim, F.H. Barido, and S.K. Lee, "Comparison of meat quality, fatty acid composition and aroma volatiles of dry-aged beef from Hanwoo cows slaughtered at 60 or 80 months old", *Asian-Australas. J. Anim. Sci.,* vol. 33, no. 1, pp. 157-165, 2020.
[http://dx.doi.org/10.5713/ajas.19.0205] [PMID: 31480197]

[8]   A. I. Carrapiso, J. F. Tejeda, J. L. Noguera, N. Ibáñez-Escriche, and E. González, "Effect of the

genetic line and oleic acid-enriched mixed diets on the subcutaneous fatty acid composition and sensory characteristics of dry-cured shoulders from Iberian pig", *Meat Sci.,* vol. 159, p. 107933, 2020. [http://dx.doi.org/10.1016/j.meatsci.2019.107933]

[9]   L. Biondi, G. Luciano, D. Cutello, A. Natalello, S. Mattioli, A. Priolo, M. Lanza, L. Morbidini, A. Gallo, and B. Valenti, "Meat quality from pigs fed tomato processing waste", *Meat Sci,* vol. 159, p. 107940, 2017.
[http://dx.doi.org/10.1016/j.meatsci.2019.107940]

[10]   L.F. Mueller, J.C.C. Balieiro, A.M. Ferrinho, T.D.S. Martins, R.R.P. da Silva Corte, T.R. de Amorim, J. de Jesus Mangini Furlan, F. Baldi, and A.S.C. Pereira, "Gender status effect on carcass and meat quality traits of feedlot Angus × Nellore cattle", *Anim. Sci. J.,* vol. 90, no. 8, pp. 1078-1089, 2019.
[http://dx.doi.org/10.1111/asj.13250] [PMID: 31240763]

[11]   G. Lombardi-Boccia, S. Lanzi, and A. Aguzzi, "Aspects of meat quality: Trace elements and B vitamins in raw and cooked meats", *J. Food Compos. Anal.,* vol. 18, no. 1, pp. 39-46, 2005.
[http://dx.doi.org/10.1016/j.jfca.2003.10.007]

[12]   OECD/FAO, "OECD-FAO Agricultural Outlook 2019-2028", *Paris/Food and Agriculture Organization of the United Nations,* OECD Publishing: Rome, 2019. [Online] Available: https://www.oecd-ilibrary.org/agriculture-and-food/oecd-fao-agricultural-outlo-k-2019-2028_agr_outlook-2019-en [Accessed Aug. 10, 2020].

[13]   USDA, "Livestock and poultry: world markets and trade", *United States Dep. Agric. Foreign Agric. Serv.,* p. 31, 2019. [Online] Available: https://usda.library.cornell.edu/ concern/publications/ 73666448x [Accessed Aug. 10, 2020].

[14]   OECD, *Per capita meat consumption by region,* 2019. [Online] Available: https://www.oecd-ilibrary.org/
search?value1=meat+per+capita+consumption&option1=quicksearch&facetOptions=51&facetNames=pub_igoId_facet&operator51=AND&option51=pub_igoId_facet&value51=%27igo%2Foecd%27&publisherId=%2Fcontent%2Figo%2Foecd. [Accessed Jul. 07, 2020].

[15]   A. Bernués, G. Ripoll, and B. Panea, "Consumer segmentation based on convenience orientation and attitudes towards quality attributes of lamb meat", *Food Qual. Prefer.,* vol. 26, pp. 211-220, 2012.
[http://dx.doi.org/10.1016/j.foodqual.2012.04.008]

[16]   R. A. Da Silva-Buzanello, D. L. Kalschne, S. M. Heinen, C. Pertum, A. F. Schuch, M. P. Corso, and C. Canan, "Pork: Profile of the West of Paraná consumers and physical evaluation of chop", *Semin. Agrar,* vol. 38, no. 6, pp. 3563-3578, 2017.
[http://dx.doi.org/10.5433/1679-0359.2017v38n6p3563]

[17]   J. D. Wood, "Meat Composition and Nutritional Value", In: *Lawrie's Meast Science,* F. Toldra, Ed., Woodhead Publishing, 2017, pp. 635-659.

[18]   M. Henchion, M. McCarthy, V.C. Resconi, and D. Troy, "Meat consumption: trends and quality matters", *Meat Sci.,* vol. 98, no. 3, pp. 561-568, 2014.
[http://dx.doi.org/10.1016/j.meatsci.2014.06.007] [PMID: 25060586]

[19]   S. Barbut, *The Science of Poultry and Meat Processing.* University of Guelph: Ontario, Canada, 2018.

[20]   L.C. Shan, A. De Brún, M. Henchion, C. Li, C. Murrin, P.G. Wall, and F.J. Monahan, "Consumer evaluations of processed meat products reformulated to be healthier - A conjoint analysis study", *Meat Sci.,* vol. 131, pp. 82-89, 2017.
[http://dx.doi.org/10.1016/j.meatsci.2017.04.239] [PMID: 28494317]

[21]   M. Faustino, M. Veiga, P. Sousa, E.M. Costa, S. Silva, and M. Pintado, "Agro-food byproducts as a new source of natural food additives", *Molecules,* vol. 24, no. 6, pp. 1-23, 2019.
[http://dx.doi.org/10.3390/molecules24061056] [PMID: 30889812]

CHAPTER 2

Dry Fermentation Technology

Abstract: The technology of the production of dry-fermented products is old and even in current days these meat products are produced, appreciated and consumed worldwide. The fermentation and drying steps are responsible for chemical and biochemical transformations that ensure the final characteristics of the product were reached. One of the disadvantages in dry-fermented product processing is the time required for all transformations that characterize the product and meets the required standard of safety and quality. Thus, the use of emerging technologies allows the improvement in the production process considering the use of new approaches, always ensuring food safety and quality. In this chapter, the ultrasound, pulsed UV light and QDS process® are discussed. The approach includes the advantages and disadvantages of each technology, the report of researchers describing the use of mentioned technologies and the most probable mechanisms associated with the effects.

Keywords: Fermented Sausages, Ham, Pulsed UV Light, Quick-Dry-Slice Process®, Salami, Starter Culture, Ultrasound.

INTRODUCTION

Dry-fermented meat products are consumed worldwide representing a part of traditional diets and perceived as attractive gastronomic entities contributing to cultural and geographic distinctiveness [1]. This processed meat product group is characterized by a ripening process accompanied by a more or less intense drying and even large of chemical and biochemical transformations [2]. Sausages and some meat cuts are the most popular dry-fermented products, which were named and prepared according to the influence of specific regions and diverse types had the designation of origin recognized. Some examples include Parma ham from Italy, Chorizo, Serrano and Iberian ham from Spain, Milan and Italian salamis, and Sobrassada from Mallorca (Islas Baleares) [2 - 5].

The dry fermented technique was mainly associated with pork meat industrialization [6].

The sausage consisting of seasoned and salted minced meat that is stuffed into intestinal casings, while is based on the art of dry ham production previously developed by the Celts and Gauls, consisting of the salting and drying of hind legs

Daneysa L. Kalschne, Marinês P. Corso & Cristiane Canan

of wild boars and pigs [1]. The most popular and famous dry-fermented cut is ham; however, the shoulder and loin are also produced and consumed.

Over the years, fermented meats came to exist in an overwhelming variety, characterized by large differences in ingredients, shape and caliber, and processing conditions. Several types of meat have been applied, including pork, beef, horse, donkey, deer, poultry, and ostrich [1]. In this context, the possibility of the use of emerging technologies on dry-fermented products represents alternatives in order to improve food safety, quality, and manufacturing process time consumption.

TRADITIONAL DRY-FERMENTED PRODUCTS ELABORATION PROCESS

Salami is prepared by grinding the meat and fat and mixing with all other ingredients including salt, curing salts as nitrite and or nitrate, sugars, spices, antioxidants, and starter culture [6]. Moreover, the processing technologies may differ depending on the drying intensity, ripening length or smoking [2]. Table **1** summarizes some dry-fermented meat product formulations. Note the high salt content used in the formulations, and considering that during the maturation stage the product can lose up to 50% moisture, which causes the salt content in the final product to increase significantly. Thus, the reduction of the salt content of these products is necessary and will be discussed later on as new technologies can contribute.

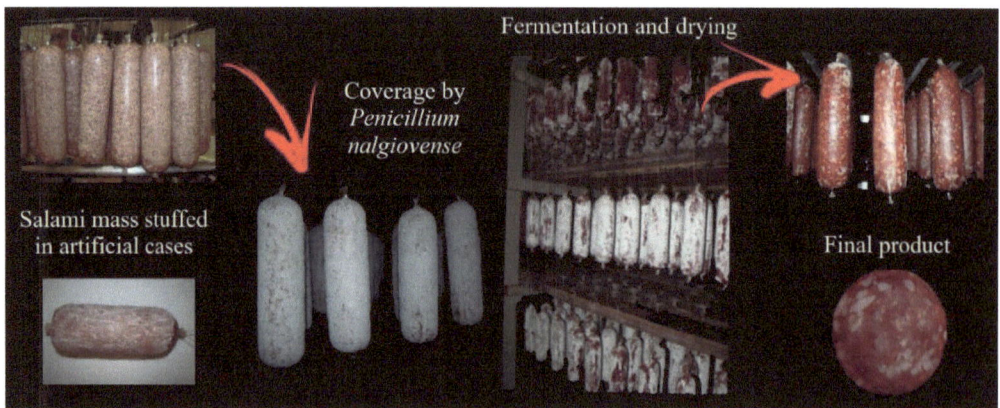

Fig. (1). Characteristics of salami during product elaboration.

The mass is stuffing in artificial casings and is submitted to a fermentation/drying for a few days with the external development of mould (*Penicillium nalgiovense*), and ripening for sufficient time up to reaching the desired water activity in the final product (Fig. **1**) [6]. The dry-fermented products were generally packed and stored under a modified atmosphere composed of nitrogen and carbon dioxide (80:20, v/v) or under vacuum to enhance their shelf life stability by minimizing microbial growth and lipid oxidation processes [7, 14].

Table 1. Dry-fermented meat products formulation.

Meat product	Meat base	Ingredients and additives	Starter culture	Reference
Italian salami	100% pork meat	1% salt, 0.3% curing salt, 1% flavoring, 1% sucrose, and 0.25% sodium erythorbate	*S. xylosus, S. carnosus, P. pentosaceus* and *P. acidilactici*	[7]
Italian salami	70% pork meat, and 30% pork fat	Salt, spices, aromas, sodium nitrate, sodium nitrite, ascorbic acid, citric acid, milk powder, lactose, and sucrose	*Lactobacillus*	[8]
Italian salami	62.09% pork meat, and 19.10% beef	2.91% salt, 0.28% sodium nitrite and nitrate, 0.23% sodium ascorbate, 0.19% pepper, 0.19% garlic powder, 0.19% nutmeg, 0.28% glucose, and 0.19% sucrose	*P. pentosaceus,* and *S. xylosus*	[9]
Milano salami	75% pork meat, and 25% belly	2.3% NaCl, 0.2% black pepper, 0.1% white pepper, 0.02% garlic, sodium nitrite, sodium ascorbate, dextrose, and saccharose	*L. sakei*	[10]
Milano salami	74.95% pork meat, and 30% pork backfat	2.5% salt, 0.3% curing salt, 0.20% sodium erythorbate, 0.02% white pepper, 0.20% garlic, 1,.0% wine, and 0.8% sugar	*D. hansenii, L. sakei, P. pentosaceus, S. carnosus, and S. xylosus*	[11]
Milano salami	69.5% pork meat, and 23% pork fat	2.3% NaCl, 0.015% sodium nitrite, 0.3% sodium erythorbate, 0.6% flavoring, 1.0% wine, 2.7% milk powder, 0.35% glucone delta lactone, 0.25% sucrose, and 0,25% glucose,	-	[12]
Chorizo	100% pork meat	2% NaCl, 75 to 150 mg/kg of sodium nitrite, and 0.5% dextrose	*L. plantarum,* and *L. delbrueckii*	[4]
Chorizo	75% pork meat, 18% pork backfat	1.5% NaCl, and 0.3% curing salt, and 5.2% sodium ascorbate	*L. plantarum*	[5]

(Table 1) cont.....

Chorizo	83% lean pork meat, and 17% fat	2.4% NaCl, 0.015% nitrite, 1.01% paprika, 0.3% aroma, 0.15% garlic, 1.7% sugars, 0.05% antioxidant, 0.745% casein, and 0.01% coloring	-	[13]
Sobrassada	16.7% lean pork meat, and 76.1% white fat	2.4% salt, and 4.8% paprika	*L. sakei*, and *S. carnosus*	[3]

BIOCHEMICAL TRANSFORMATIONS

The fermentation process is based on the anaerobic metabolism of carbohydrates by microorganisms. Starter cultures were generally used for food preservation and the development of sensory characteristics, nutritional value and food safety ensure [15]. During the fermentation and ripening processes, the sugar added in the salami or pork cut is fermented by *Lactobacillus* and converted into organic acids, resulting in pH reduction, reaching the isoelectric point of meat proteins and promoting myofibrillar proteins insolubilization and the consequent water loss by dehydration [16]. The relative humidity and temperature were crucial controllable parameters for suitable fermentation and ripening process. In general, the relative humidity and temperature should be decreased with the advance of ripening time. Table **2** summarizes some dehydration processes employed for fermented products. Between 24 and 48 h, high moisture and temperature step are generally performed, and subsequently, more pronounced drying steps (lower relative humidity and temperature) are applied up to the desired water activity and/or moisture were reached in the final product. The ripening time is dependent on some factors, such as chamber relative humidity, temperature, and air circulation, casing caliber and composition, formulation, and the final product water activity and moisture desired [5, 10, 17]. The weight loss increases during ripening as a consequence of the drying process in dry-fermented products. Values of weight loss from 30% to 40% for Italian salami [7], from 32.13% to 41.2% for Milano salami [11, 12], and from 27.57% to 41.74% for chorizo [5, 19] were reported in the literature. The variability in weight loss was also associated with the same factors previously mentioned for ripening time.

Beside ripening, the weight loss and structure of dry and fermented products are largely associated with the starter cultures action in the product, which induces a decrease in pH from 5.8 to 5.0, whereas water simultaneously diffuses from the meat matrix into the surrounding phase upon drying. The weight loss and pH decrease are indispensable processes for the formation of a sliceable meat product such as dry fermented sausages and cuts [14].

Table 2. Dehydration process applied for dry-fermented meat products.

Meat product	Caliber (mm)	Relative humidity (%)/Temperature (°C)								Ripening (days)	References
		1st day	2nd day	3rd day	4th day	5th day	6th day	7th day	8th day*		
Italian salami	40	70-99/18-22	70-99/18-22	70-99/18-22	71-76/12-16	71-76/12-16	71-76/12-16	75-80/10-12	75-80/10-12	-	[8]
Italian salami	60	95/25	92/24	89/23	86/22	83/21	80/20	80/19	75/18	28	[19]
Italian salami	60	95/25	93/24	90/23	85/22	80/21	75/20	75/18	75/18	28	[9]
Milano salami	40	95/25	93/24	90/23	85/22	80/21	75/20	75/18	75/18	42	[20]
Milano salami	45	90/25	90/25	83/15	83/15	83/15	83/15	83/15	83/15	26	[12]
Chorizo	40	90/23	90/23	75/13	75/13	75/13	75/13	75/13	75/13	20	[5]
Chorizo	32-34	85/7	85/7	75-80/12	75-80/12	75-80/12	75-80/12	75-80/12	75-80/12	-	[21]
Chorizo	-	90/5	80/4-7	80/4-7	80/4-7	80/4-7	8/75	8/75	8/75	-	[18]

* In the subsequent days the fermented products were kept in the same conditions of 8th day up to the end of ripening.

Fig. (2) presents the relationships between weight loss and pH decrease during ripening; these are real examples from the authors' experience that was not published but clearly describe the phenomena. Furthermore, acid pH is one of the main physicochemical and sensorial characteristics of salami. Acid pH also acts as a barrier to the development of undesirable microorganisms in the product, especially when associated with low water activity [16]. The stability of salami was ensured by barriers such as acid pH, lower water activity, and lower sugars content after fermentation step.

USE OF NEW TECHNOLOGIES IN DRY-FERMENTED PRODUCTS

Novel and emerging food processing technologies for fermentation have increasingly gained interest over the past years [22]. These new technologies represent a considerable improvement in food safety and quality and the reduction in manufacturing time for dry-fermented products. Three emerging technologies applied for dry-fermented products were discussed below, the ultrasound, pulsed UV light and Quick-dry-slice process®.

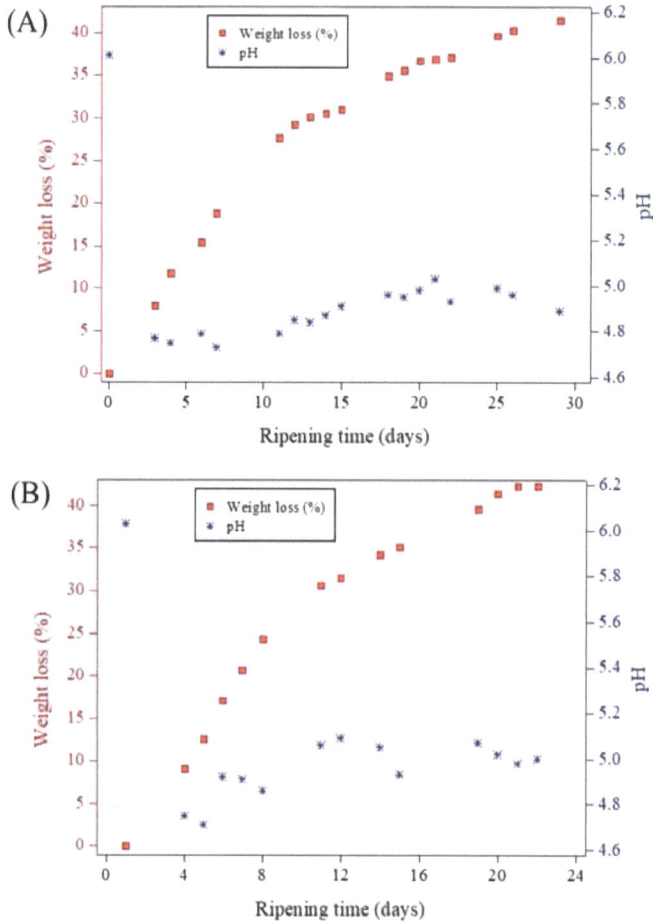

Fig. (2). Dry-fermented products weight loss and pH decrease during the ripening process: **(A)** Milano salami; **(B)** Italian salami.

ULTRASOUND

Ultrasound technology is based on the mechanical waves at a frequency which is above the threshold of human hearing (above 16 kHz) and its effect on the biological materials are generally divided into two categories: thermal and nonthermal mechanisms [23]. Ultrasound is one of the emerging technologies that were developed to minimize processing, maximize quality and ensure the safety of food products [24].

Low energy (low power, low intensity) ultrasound has frequencies higher than 100 kHz at intensities below 1 W cm^{-2} and can be used for non-invasive analysis and monitoring of food materials both during processing and storage to ensure high quality and safety. In contrast, the high energy (high power, high-intensity) ultrasound with intensities higher than 1 W cm^{-2} at frequencies ranging from 20 to 500 kHz can disrupt and induce effects on the physical, mechanical or chemical/biochemical properties of foods [24]. These high energy ultrasound effects are promising in food processing, preservation and safety. Furthermore, the literature available extensively demonstrates the potential of US against foodborne pathogens and enzymes [24, 26]. Regarding the ultrasound parameters and units, the frequency (kHz), power (W), exposure time (min), and temperature (°C) are commonly used to characterize the treatment.

A study applied the ultrasound in Italian salami previously vacuum packaged in polyethylene bags - to avoid the water absorption - in an ultrasonic bath at 20 °C, 25 kHz and a power of 128 W (for 3 min exposure) [7]. Each batch was sonicated for 0 (control), 3, 6 or 9 min and the salamis were disposed of in a maturation chamber for 28 days, vacuum packaged and storage at room temperature for 120 days. The results demonstrated that ultrasound exposure on the first day of maturation improved the growth of lactic acid bacteria.

The versatility of ultrasound on the stimulation and retardation of lactic acid bacteria and on the potential that exists to apply it to food fermented products allows expanding the range of possibilities [25]. The authors [7] also reported that after 15 days of maturation, the treatments with 6 min and 9 min of exposure had higher counts than the control for lactic acid bacteria and *Staphylococcus*, pointed out that microorganisms surviving the sonication kept their microscopic appearance and growth rate unchanged.

Moreover, the effect of ultrasound (20 kHz, power of 12 W during 20 min) in inhibition and cell damage in *E. aerogenes*, *Bacillus subtilis* and *Staphylococcus* spp. was evaluated. It was observed that for *E. aerogenes* and *B. subtilis* lethal damage reach up to 4.5-log reduction, while the treatment did not affect *Staphylococcus* spp [26]. According to the authors, the main reason for bacterial resistance to ultrasonic deactivation is the characteristic of the bacterial capsule, considering that microorganisms with a thicker and "soft" capsule are highly resistant to the ultrasonic process.

Some hypotheses were mentioned [27] that could justify the Gram-positive cocci *Staphylococcus* spp. resistance. Larger cells are more sensitive than the small ones, probably due to its larger surface area; thus, cocci are more resistant than bacilli due to the relationship of cell surface and volume. The Gram-positive bacteria are more resistant than gram-negative ones, possibly because of their thicker cell wall which provides them better protection against ultrasound effects. Finally, spores are very hard to destroy compared to vegetative cells, which are in the phase of growth.

Regarding the salami produced [7] and storage for 120 days, the treatment of 9 min of ultrasonic exposure presented higher lactic acid bacteria counts, showing that it is the most effective exposure time for microbial growth and/or maintenance of viability of fermentative microorganism in this study. The authors reinforced that the delayed effect is possibly due to the effect of ultrasound on the breakage of macromolecules and the consequent increase in nutrient availability and that the ultrasound reduced the competitive microorganisms over time [28].

Similarly, the authors [25] reported that a higher specific growth rate and a shorter lag phase were observed for *Lactobacillus* sakei in a model system at low ultrasound power (2.99 W, 20 kHz, 5 min) compared to control (without ultrasound). Conversely, a decrease in specific growth rate and an increase in the lag phase were observed with an increase in the ultrasonic power level (68.5 W, 20 kHz). Microorganisms inactivation tests are generally conducted at 20 kHz, a frequency at the low end of the power ultrasound frequency spectrum [28].

In general, some hypotheses are mentioned [7, 27 - 29] and detailed in Fig. (**3**) for microorganisms ultrasound stimulation effect.

In contrast with the range of possibilities of ultrasound exposure regarding microbial growth rate, sonication in the presence of oxygen leads to the generation of oxygen atoms, which reacts with hydrogen atoms to form a peroxide radical (HOO•). Some chemical interactions between HOO• and organic substances in solution or biomacromolecules lead to changes in the properties of these compounds or changes in biomolecules [23].

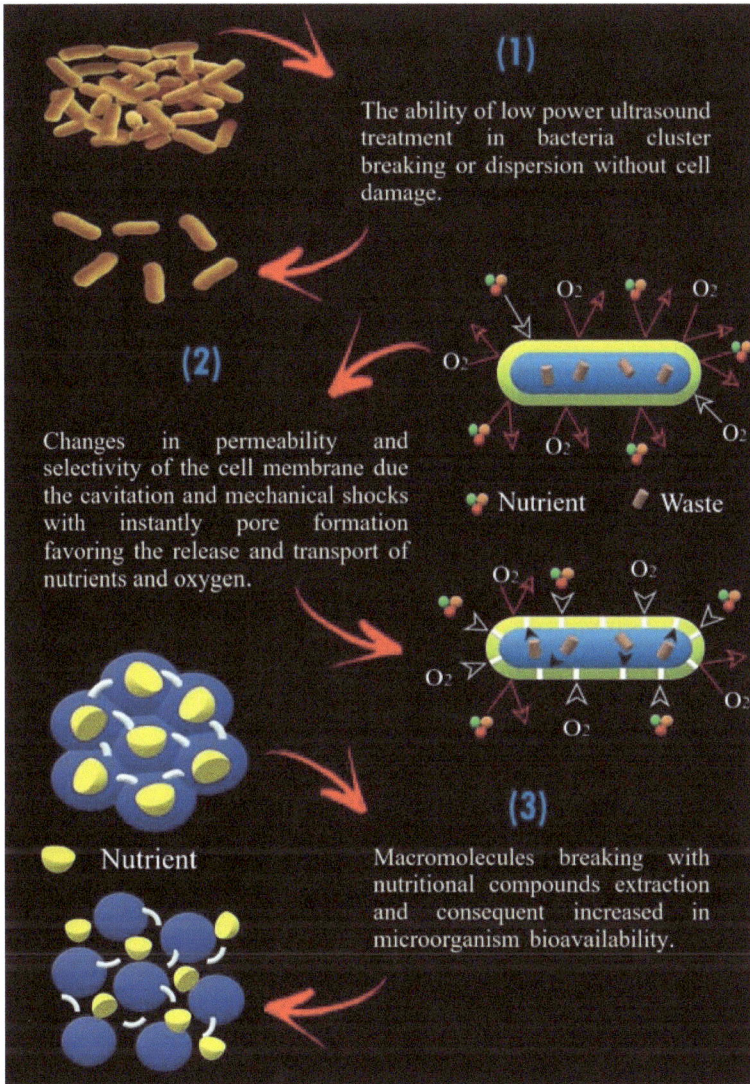

Fig. (3). Hypothesis for microorganism stimulation by ultrasound.

The thiobarbituric acid reactive substances (TBARS) are considered the major oxidation products of poly unsaturated fatty acid (PUFA) during meat product storage. Thus, the TBARS values (mg malondialdehyde-MDA kg^{-1}) was employed as a lipid oxidation index for meat and meat products [23]. In this context, the malondialdehyde (MDA) production evoked by ultrasonication is comparable to that initiated by iron ions, and the cavitation seems responsible for free radicals arising [29]. The ultrasound frequency, power level and processing time are described as variables that play an influence on lipid oxidation [30].

Although it has been reported that sonochemical effects on oxidation are more evident at higher frequencies (> 200 kHz) because more active bubbles are formed, a study [7] reported an acceleration of lipid oxidation even in lower frequencies (25 kHz).

The mechanical effects of ultrasound provide greater penetration of solvent into cellular materials and improve mass transfer in organic compounds extraction from a tissue [27]. If the opposite way is considered, ultrasound can be employed to increase the rate of heat transfer between a solid heated surface and a liquid, a process of critical importance in drying [31]. For dry-fermented products, two steps are critical. Fermentation provides pH reduction up to the isoelectric point of meat protein, and this physico-chemical change allows further drying in order to complete the product production process [16]. On cured meat, ultrasound has been considered as an emerging technology, suggesting that micro-channels, micro-agitation and free radicals generated by cavitation contribute accelerating mass transport and structural changes. The possibility of an ultrasound application could be used to intensify the curing of dry meats and could improve the texture, surpassing the traditional drying process - considered usually a prolonged and energy-intensive process [32].

No study in the literature available reports the ultrasound effects in sensory acceptance of dry-fermented products. However, due to the direct relationship between the increase in oxidation rate and ultrasound exposure, some aroma and flavor changes will be expected in sensory acceptance [29]. The threshold for MDA was in the range between 0.3 and 2 mg MDA kg^{-1} and some studies reported MDA content above the threshold [7, 24].

PULSED UV LIGHT

The emergent technology of pulsed UV irradiation originated from xenon lamps pulsed irradiation studies that reported microorganisms inhibition for food decontamination and water disinfection [33]. The classical UV-C treatment works in a continuous mode with the use of monochromatic light (254 nm) produced by low-pressure a mercury lamp. With the advancement of technical evolution, xenon lamps that can produce flashes several times per second were used in pulsed light emergent technology. The pulsed light is a technique for surface decontamination by microorganisms inactivation using the emission of short-time pulses of an intense broad spectrum, rich in UV-C light (between 200 and 280 nm); although the range from 200 to 1100 nm can be used [34]. The following parameters and units are commonly used to characterize a pulsed light treatment: fluence (J/m^2 or J/cm^2), exposure time (s); pulse width (fractions of s), pulses per second (pps), and peak power (W).

Regarding the use of continuous or pulsed UV light effects, the operational differences were based on the fact that in the pulsed the power is magnified by storing electricity in a capacitor over relatively long times (fractions of a second) and releasing it in a short time (millionths or thousandths of a second) [34].

A study evaluated the continuous *versus* pulsed UV-LED in *Bacillus globiggi* spores inactivation [35]. The authors achieve a 6-log inactivation at a fluence of 665 J/m^2 for conventional exposure while for pulsed only a fluence of 365 J/m^2 was required. In this context, pulsed UV-LED was considered more efficient than the conventional approach. Studying bacterial vegetative cells, it was reported a similar inactivation for coliform and *E. coli* evaluated in a buffered lab employing a 265 nm UV-LED (light-emitting diodes) continuous and pulsed irradiation (both at equivalent UV dose 4.9 mJ cm^2) [33]. However, it is important to mention that the cited authors extend the exposure time to pulsed irradiation in order to achieve equivalent UV fluence as those of continuous irradiation.

Some researchers pointed out that short-duration of pulsed caused more damage in cell structure if compared with the correspondent fluence rate of conventional exposure [35]. Furthermore, two reasons were mentioned: (1) the LED photon generation is increased in proportion to the sharp increase in current that is applied during pulsing, thus pulsed peak photon discharge is higher than conventional; (2) refers to the broad wavelength spectrum, which induces both visible and infrared regions with biocide potency.

Certainly, the use of pulsed light has shown great potential for application in surface food products with small thickness. A study was conducted by employing pulsed UV light at a fluence of 3 J/cm^2 (1 pulse) in order to reduce the surface contamination of fermented salami slices [36]. The results demonstrated around 2-log reduction for *Listeria monocytogenes*, *Escherichia coli* 0157:H7, *Salmonella* Typhimurium and *Staphylococcus aureus;* the fluence increase up to 15 J/cm^2 (5 pulses; one pulse per 2 s) did not improve the inhibition rate. Moreover, it was also demonstrated that pulsed light is effective to reduce *L. monocytogenes* and *Salmonella* Typhimurium on the surface of ready-to-eat dry cured meat products in the range of 1.5 to 1.8 log CFU/cm^2 at a fluence of 11.9 J/cm^2 [37]. However, the authors evaluated a fluence interval from 0.7 to 11.9 J/cm^2 and in the range studied (each pulse is delivered in 250 μs and corresponds to a fluence of 0.7 J/cm^2), an increase in the fluence demonstrated to improve the bacteria inhibition.

Regarding the yeast inhibition, after two flashes of pulsed light irradiation (two flashes at 0.7 J/cm^2/flash; total: 1.4 J/cm^2), distinct structural changes in the yeast cells were observed [38].

It is important to emphasize that microorganisms located on the upper layers receive light directly and are easily inactivated. However, bacteria disposed of in the bottom layers are less affected, because they are protected by those on the upper layers which screen the light and allow survival of shadowed cells [36]. According to Mahendran *et al.* [39], the pulsed light mechanism of action involves the absorption of UV light by microbial DNA, considering that the light leads physico-chemical changes in its structure, resulting in damage of genetic information, impaired replication and gene transcription as well as the eventual death of the cell. Moreover, the pulsed light mechanisms are associated with the photochemical reactions that damage the nucleic acid or the photothermal effects that induce membrane disruption and protein unfolding [35].

The continuous UV light induces the lipid peroxidation and it probably occurs due to the long light exposure time for microbial inactivation on the meat surface. Pulsed light is very intensive UV light and the lipid peroxidation in treated meat samples may also occur, however at a smaller rate [34]. Rajkovic *et al.* [36] reported both the increase in lipid and protein oxidation, evidenced by MDA and protein carbonyl higher content on samples treated with pulsed light, notably at a fluence of 15 J/cm^2.

Depending on the pulsed light treatment applied, some sensory undesirable changes were described. Tomašević [40] evaluated the impact of intense pulsed light in dry ham and fermented sausage by sensory test. Light pulses with a duration of 300 μs, pulse intensity of 3.4 J/cm^2 was applied considering 1 pulse and 5 pulses at a rate of one pulse per 2 seconds. Eight trained assessors evaluated the appearance, color, odor and taste, and texture and juiciness using a 5-point scale (5 = excellent, typical quality, without visible defects; 3 = neither good nor poor quality, still can be used for its intended purpose; and 1 = unacceptable, extremely poor quality, cannot be used for its intended purpose). Both meat products showed greater resistance to the impacts of pulsed light. The authors reported no differences for all attributes evaluated between 1 pulse treated and control samples of Parma ham and sausage ($p > 0.05$). When the higher fluence of 17 J/cm^2 (5 – pulses treatment) was applied, the ham and fermented sausage texture and juiciness, and the Parma ham odor and taste significantly decreased ($p < 0.05$) to the level of neither good nor poor.

The color, odor and flavor of dry-fermented *salchichón* had no significant differences ($p > 0.05$) even treated at different fluences (from 0 to 11.9 J/cm^2) after 0, 15, and 30 days of storage. The samples were evaluated by 12 trained assessors using a 6-point scale (in which 1 corresponded to the lowest preference and 6 to the highest preference) [37].

QUICK-DRY-SLICE PROCESS®

In the productive process of dry-fermented meat products, the ripening is the most time and energy consuming step as discussed in Table **2**, which implies in a few weeks up to years to reach the desired product characteristics and requires a considerable investment [41]. On fermentation and drying steps, the meat product is dehydrated and some biochemical reactions produced by endogenous or microbial enzymes degraded lipids and proteins and contribute to confer the characteristic texture and flavor [42]. However, the author also emphasized that in conventional production processes, some deviation could affect the final characteristic of dry-fermented meat products. In order to solve this question, the authors [43] proposed and patented a new technique called Quick-Dry-Slice® (QDS®). This technology accelerates the drying/ripening process of meat products by the use of convective air in a continuous system. The production of dry-fermented sausages through QDS process® implies fermentation to the desired pH, freezing, slicing and drying in a continuous system for a short time even at product temperature below 18 °C [13].

Table **3** summarizes the literature information's regarding the application of QDS process® in dry-fermented products. The following parameters and units are commonly used to characterize a QDS process®: relative humidity (%), pressure (mbar), chamber temperature (°C), and exposure time (min). A previous dry step can be performed for some products if necessary. Relative humidity employed is generally lower in order to allow the water migration from the product and as temperature employed is considered lower, it is necessary to use a pressurized chamber. Exposure time is correlated with the product weight loss desired.

Samples of *salchichón* produced by conventional and QDS process® were evaluated regarding the analysis of pH, *Staphylococcus aureus*, sulphite reducing *Clostridium*, *Escherichia coli*, *Listeria monocytogenes*, and *Salmonella* after 15 days at 4 °C, 3 months at 1 °C and 13 °C. The results indicated that the pH of the QDS process® was higher than the conventional ones. However, both processes reduced the count of *Staphylococcus aureus*, sulphite reducing *Clostridium*, and *Escherichia coli* after processing and storage to similar and acceptable values. For *Salmonella,* if it is present in the raw material it was detected in the final product, independent of the applied process, and *Listeria monocytogenes* was below the limit of detection in all samples [42].

Comparing the chorizo produced by conventional (28% weight loss) and QDS process® (27, 30 and 33% weight loss), the attributes of color intensity and hardness were higher for QDS process®, while the acidity flavor is similar for all samples, and the chorizo flavor intensity was higher for conventional sample [46].

In a different approach, Ferrini *et al.* [44] evaluated the physico-chemical and sensory parameters of salami and chorizo using QDS process® preceded by a drying step at 53 °C for 0, 50, 65, 80, 95 and 110 min. Regarding sensory evaluation, the cooked appearance and cooked fat odor, and coked flavor (all undesirable) of both products became higher ($p < 0.05$) than in the conventional samples. In contrast, cohesiveness and hardness are not influenced in each group of samples even considering the previous step of drying. The water activity was not affected by the different thermal processing times applied before QDS process® for both salami and chorizo, while the pH was not affected only for salami. In contrast, the lightness, redness and yellowness of the salami and chorizo decreased with an increase in the previous step of drying (0 to 110 min).

Table 3. QDS process® applied for meat products elaboration.

| Meat product | Previous dry step | QDS process®applied | | | | | | References |
		Relative humidity	Air velocity	Pressure	Chamber/ slices temperature	Exposure time	Product weight loss	
Ham	5 °C/25 days or 12 °C/8 days	40%	-	-	30 °C/ 20 °C	50 min	-	[41]
Restructured minced	-	-	-	40 mbar	- /8 °C	40 min	30%	[44]
Fish sausages	-	35-40%	3.5 m s^{-1}	-	30 °C/ -	45 min	40%	[45]
Chorizo	53 °C/ 65 min	-	-	40 mbar	43 °C/ -	45 min	29%	[13]
Chorizo	-	30%	-	-	30 °C/ -	33-43 min	27-33%	[46]
Salami/ chorizo	53 °C/0-110 min	-	-	40 mbar	- /14	38 min	30%	[47]
Milano salami	-	50-60%	-	-	20-25 °C/ -	50 min	40%	[10]

The innovative QDS process® facilitates the reduction of the drying period of sliced products by direct drying of slices in a continuous system. Additionally, better control of processing and product quality provides great flexibility in production planning, an aspect which is of great importance for product development [41]. Furthermore, the results described for microbiological and sensory quality ensure the suitability and safe approach of QDS process®.

FINAL CONSIDERATION

The emerging technologies presented in this chapter stand out as significant advances in dry-fermented product processing technology. Ultrasound has been used on a larger scale in the studies described in the literature, mainly showing effects on the stimulation or reduction of fermentative, deteriorating and pathogenic microorganisms. Effects on lipid oxidation and favoring the drying of the product have also been reported. The pulsed light has been applied with a focus on the microbial inhibition of fungi, bacteria and spores in slices of dry-fermented products since light does not exceed structures of thicker thickness. QDS process® is a technology that arises as an option for reducing processing time, ensuring food safety and dry-fermented product quality, proving to be an alternative technology in terms of processing.

REFERENCES

[1]   F. Leroy, and L. De Vuyst, "Fermented Foods: Fermented Meat Products", In: *Encyclopedia of Food and Health,* B. Caballero, P.M. Finglas, F. Toldrá, Eds., Academic Press: London, 2016, pp. 656-660.

[2]   F. Toldrá, "Process Improving the sensory quality of cured and fermented meat products", In: *Processed Meats: Improving Safety, Nutrition and Quality.,* J.P. Kerry, J.F. Kerry, Eds., Woodhead Publishing: Oxford, 2011, pp. 508-526.
[http://dx.doi.org/10.1533/9780857092946.3.508]

[3]   P. Llull, S. Simal, A. Femenia, J. Benedito, and C. Rosselló, "The use of ultrasound velocity measurement to evaluate the textural properties of sobrassada from Mallorca", *J. Food Eng.,* vol. 52, no. 4, pp. 323-330, 2002.
[http://dx.doi.org/10.1016/S0260-8774(01)00122-4]

[4]   I. Nikodinoska, L. Baffonia, D. Gioia, B. Manso, L. García-Sánchez, B. Melero, and J. Rovira, "Protective cultures against foodborne pathogens in a nitrite reduced fermented meat product", *Lwt,* vol. 101, pp. 293-299, 2019.
[http://dx.doi.org/10.1016/j.lwt.2018.11.022]

[5]   C.P. Cavalheiro, C. Ruiz-Capillas, A.M. Herrero, F. Jiménez-Colmenero, T. Pintado, C.R. de Menezes, and L.L.M. Fries, "Effect of different strategies of *Lactobacillus plantarum* incorporation in chorizo sausages", *J. Sci. Food Agric.,* vol. 99, no. 15, pp. 6706-6712, 2019.
[http://dx.doi.org/10.1002/jsfa.9952] [PMID: 31350779]

[6]   C. Alamprese, L. Fongaro, and E. Casiraghi, "Effect of fresh pork meat conditioning on quality characteristics of salami", *Meat Sci.,* vol. 119, pp. 193-198, 2016.
[http://dx.doi.org/10.1016/j.meatsci.2016.05.004] [PMID: 27214278]

[7]   L.L. Alves, "Effect of ultrasound on the physicochemical and microbiological characteristics of Italian salami", *Food Res. Int.,* vol. 106, pp. 363-373, 2018.
[http://dx.doi.org/10.1016/j.foodres.2017.12.074]

[8]   S. Petrini, F. Feliziani, C. Casciari, M. Giammarioli, C. Torresi, and G.M. De Mia, "Survival of African swine fever virus (ASFV) in various traditional Italian dry-cured meat products", *Prev. Vet. Med.,* vol. 162, pp. 126-130, 2019.
[http://dx.doi.org/10.1016/j.prevetmed.2018.11.013] [PMID: 30621891]

[9]   A.M. Backes, C.P. Cavalheiro, F.S. Stefanello, F.L. Lüdtke, N.N. Terra, and L.L.M. Fries, "Chemical composition, microbiological properties, and fatty acid profile of Italian-type salami with pork backfat substituted by emulsified canola oil", *Cienc. Rural,* vol. 47, no. 8, pp. 1-7, 2017.
[http://dx.doi.org/10.1590/0103-8478cr20160688]

[10] M.N. Haouet, "Evaluation of the safety of Milano-type dry fermented sausages produced by a fast drying technology", *Ital. J. Food Sci.,* vol. 29, no. 3, pp. 550-558, 2017.

[11] E. P. Krummenauer, G. O. Paranhos, J. F. Silva, R. A. Silva-Buzanello, D. L. Kalschne, M. P. Corso, and C. Canan, "Salame tipo Milano com substituição parcial do toucinho por queijo mussarela", *Rev. Cultiv. Saber,* vol. 8, no. 2, pp. 143-161, 2015.

[12] A.C.G. Mendes, and D.M. Rettore, "A. de L. S. Ramos, S. de F. V. da Cunha, L. C. de Oliveira, and E. M. Ramos, "Salames tipo milano elaborados com fibras de subprodutos da produção de vinho tinto", *Cienc. Rural,* vol. 44, no. 7, pp. 1291-1296, 2014.
[http://dx.doi.org/10.1590/0103-8478cr20130389]

[13] K. Stollewerk, A. Jofré, J. Comaposada, G. Ferrini, and M. Garriga, "Ensuring food safety by an innovative fermented sausage manufacturing system", *Food Control,* vol. 22, no. 12, pp. 1984-1991, 2011.
[http://dx.doi.org/10.1016/j.foodcont.2011.05.016]

[14] F.H. Walz, M. Gibis, K. Herrmann, and J. Weiss, "Impact of smoking on efflorescence formation on dry-fermented sausages", *Food Struct.,* vol. 20, pp. 1-8, 2019.
[http://dx.doi.org/10.1016/j.foostr.2019.100111]

[15] P.M. Malo, and E.A. Urquhart, "Fermented Foods: Use of Starter Cultures", In: *Encyclopedia of Food and Health.,* B. Caballero, P.M. Finglas, F. Toldrá, Eds., Academic Press: London, 2016, pp. 681-685.

[16] D.L. Kalschne, R.A. Silva-Buzanello, M.P. Corso, E. Colla, and C. Canan, "Microencapsulated lactic acid bacteria in meat products", In: *The Many Benefits of Lactic Acid Bacteria,* J.G. LeBlanck, A.M. LeBlanck, Eds., LeBlanck, 2019, pp. 55-73.

[17] A. Romano, P. Masi, and S. Cavella, "Visual evaluation of sliced Italian salami by image analysis", *Food Sci. Nutr.,* vol. 6, no. 1, pp. 153-159, 2017.
[http://dx.doi.org/10.1002/fsn3.540] [PMID: 29387373]

[18] C. Pereira, "Type of paprika as a critical quality factor in Iberian chorizo sausage manufacture", *CYTA J. Food,* vol. 17, no. 1, pp. 907-916, 2019.
[http://dx.doi.org/10.1080/19476337.2019.1668861]

[19] L. L. Alves, M. S. Silva, D.R.M. Flores, D.R. Athayde, A.R. Ruviaro, D.S. Brum, V.S.F. Batista, R.O. Mello, C.R. Menezes, P.C.B. Campagnol, R. Wagner, J.S. Barin, and A.J. Cichoski, "Effect of ultrasound on the physicochemical and microbiological characteristics of Italian salami", *Food Res. Int.,* vol. 106, pp. 363-373, 2018.
[http://dx.doi.org/10.1016/j.foodres.2017.12.074]

[20] M.C. Sawitzki, Â.M. Fiorentini, A.C. Junior, T.M. Bertol, and E.S. Sant'Anna, *"Lactobacillus plantarum* AJ2 isolated from naturally fermented sausage and its effects on the technological properties of Milano-type salami", *Food Sci. Technol. (Campinas),* vol. 28, no. 3, pp. 709-717, 2008.
[http://dx.doi.org/10.1590/S0101-20612008000300030]

[21] R. Thirumdas, "Evaluating the impact of vegetal and microalgae protein sources on proximate composition, amino acid profile, and physicochemical properties of fermented Spanish 'chorizo' sausages", *J. Food Process. Preserv.,* vol. 42, no. 11, pp. 1-8, 2018.
[http://dx.doi.org/10.1111/jfpp.13817]

[22] O.A. Adebo, P.B. Njobeh, A.S. Adeboye, J.A. Adebi, S.S. Sobowale, O.M. Ogundele, and E. Kayitesi, "Advances in Fermentation Technology for Novel Food Products", In: *Innovations in Technologies for Fermented Food and Beverage Industries,* S.K. Panda, P.H. Shetty, Eds., Springer: New York, 2018, pp. 71-87.

[23] K.M. Wójciak, D.M. Stasiak, K. Ferysiuk, and E. Solska, "The influence of sonication on the oxidative stability and nutritional value of organic dry-fermented beef", *Meat Sci.,* vol. 148, no. July, pp. 113-119, 2019.
[http://dx.doi.org/10.1016/j.meatsci.2018.10.010]

[24] T.S. Awad, H.A. Moharram, O.E. Shaltout, D. Asker, and M.M. Youssef, "Applications of ultrasound in analysis, processing and quality control of food: A review", *Food Res. Int.,* vol. 48, no. 2, pp. 410-427, 2012.
[http://dx.doi.org/10.1016/j.foodres.2012.05.004]

[25] K.S. Ojha, J.P. Kerry, C. Alvarez, D. Walsh, and B.K. Tiwari, "Effect of high intensity ultrasound on the fermentation profile of *Lactobacillus* sakei in a meat model system", *Ultrason. Sonochem.,* vol. 31, pp. 539-545, 2016.
[http://dx.doi.org/10.1016/j.ultsonch.2016.01.001] [PMID: 26964981]

[26] S. Gao, G.D. Lewis, M. Ashokkumar, and Y. Hemar, "Inactivation of microorganisms by low-frequency high-power ultrasound: 1. Effect of growth phase and capsule properties of the bacteria", *Ultrason. Sonochem.,* vol. 21, no. 1, pp. 446-453, 2014.
[http://dx.doi.org/10.1016/j.ultsonch.2013.06.006] [PMID: 23835398]

[27] F. Chemat, Zill-e-Huma, and M.K. Khan, "Applications of ultrasound in food technology: Processing, preservation and extraction", *Ultrason. Sonochem.,* vol. 18, no. 4, pp. 813-835, 2011.
[http://dx.doi.org/10.1016/j.ultsonch.2010.11.023] [PMID: 21216174]

[28] H. Feng, W. Yang, and T. Hielscher, "Power ultrasound", *Food Sci. Technol. Int.,* vol. 14, no. 5, pp. 433-436, 2008.
[http://dx.doi.org/10.1177/1082013208098814]

[29] P.K. Hristov, L.A. Petrov, and E.M. Russanov, "Lipid peroxidation induced by ultrasonication in Ehrlich ascitic tumor cells", *Cancer Lett.,* vol. 121, no. 1, pp. 7-10, 1997.
[http://dx.doi.org/10.1016/S0304-3835(97)00318-2] [PMID: 9459167]

[30] P. Juliano, A.E. Torkamani, T. Leong, V. Kolb, P. Watkins, S. Ajlouni, and T.K. Singh, "Lipid oxidation volatiles absent in milk after selected ultrasound processing", *Ultrason. Sonochem.,* vol. 21, no. 6, pp. 2165-2175, 2014.
[http://dx.doi.org/10.1016/j.ultsonch.2014.03.001] [PMID: 24704065]

[31] T.J. Mason, and L. Paniwnyk, L. J. P., "The uses of ultrasound in food processing", *Ultrason. Sonochem,* vol. 3, pp. S253-S260, 1996.
[http://dx.doi.org/10.1016/S1569-2868(96)80007-6]

[32] A.D. Alarcon-Rojo, H. Janacua, J.C. Rodriguez, L. Paniwnyk, and T.J. Mason, "Power ultrasound in meat processing", *Meat Sci.,* vol. 107, pp. 86-93, 2015.
[http://dx.doi.org/10.1016/j.meatsci.2015.04.015] [PMID: 25974043]

[33] K. Song, F. Taghipour, and M. Mohseni, "Microorganisms inactivation by continuous and pulsed irradiation of ultraviolet light-emitting diodes (UV-LEDs)", *Chem. Eng. J.,* vol. 343, pp. 362-370, 2018.
[http://dx.doi.org/10.1016/j.cej.2018.03.020]

[34] V.M. Gómez-López, P. Ragaert, J. Debevere, and F. Devlieghere, "Pulsed light for food decontamination: a review", *Trends Food Sci. Technol.,* vol. 18, no. 9, pp. 464-473, 2007.
[http://dx.doi.org/10.1016/j.tifs.2007.03.010]

[35] T. Tran, L. Racz, M.R. Grimaila, M. Miller, and W.F. Harper Jr, "Comparison of continuous *versus* pulsed ultraviolet light emitting diode use for the inactivation of Bacillus globigii spores", *Water Sci. Technol.,* vol. 70, no. 9, pp. 1473-1480, 2014.
[http://dx.doi.org/10.2166/wst.2014.395] [PMID: 25401310]

[36] A. Rajkovic, I. Tomasevic, B. De Meulenaer, and F. Devlieghere, "The effect of pulsed UV light on *Escherichia coli* O157:H7, *Listeria monocytogenes, Salmonella* Typhimurium, *Staphylococcus aureus* and *staphylococcal enterotoxin* A on sliced fermented salami and its chemical quality", *Food Control,* vol. 73, pp. 829-837, 2017.
[http://dx.doi.org/10.1016/j.foodcont.2016.09.029]

[37] M. Ganan, E. Hierro, X.F. Hospital, E. Barroso, and M. Fernández, "Use of pulsed light to increase the

safety of ready-to-eat cured meat products", *Food Control,* vol. 32, no. 2, pp. 512-517, 2013.
[http://dx.doi.org/10.1016/j.foodcont.2013.01.022]

[38] K. Takeshita, J. Shibato, T. Sameshima, S. Fukunaga, S. Isobe, K. Arihara, and M. Itoh, "Damage of yeast cells induced by pulsed light irradiation", *Int. J. Food Microbiol.,* vol. 85, no. 1-2, pp. 151-158, 2003.
[http://dx.doi.org/10.1016/S0168-1605(02)00509-3] [PMID: 12810279]

[39] R. Mahendran, "Recent advances in the application of pulsed light processing for improving food safety and increasing shelf life", *Trends Food Sci. Technol.,* vol. 88, pp. 67-79, 2019.
[http://dx.doi.org/10.1016/j.tifs.2019.03.010]

[40] I. Tomašević, "Intense light pulses upset the sensory quality of meat products", *Tehnol. Mesa,* vol. 56, no. 1, pp. 1-7, 2015.
[http://dx.doi.org/10.5937/tehmesa1501001T]

[41] K. Stollewerk, A. Jofré, J. Comaposada, J. Arnau, and M. Garriga, "NaCl-free processing, acidification, smoking and high pressure: Effects on growth of Listeria monocytogenes and Salmonella enterica in QDS processed® dry-cured ham", *Food Control,* vol. 35, no. 1, pp. 56-64, 2014.
[http://dx.doi.org/10.1016/j.foodcont.2013.06.032]

[42] J. Comaposada, J. Arnau, M. Garriga, M. Xargayó, J. Bernardo, and M. Corominas, "Secado rápido de productos cárnicos crudos curados. Tecnología Quick-Dry-Slice process (QDS process)", *Eurocarne,* vol. 157, pp. 1-6, 2007.

[43] J. Comaposada, J. Arnau, P. Gou, and J. M. Monfort, *"Accelerated method for drying and maturing sliced food products",* WO2005092109A1, 2004.

[44] G. Ferrini, J. Comaposada, J. Arnau, and P. Gou, "Colour modification in a cured meat model dried by Quick-Dry-Slice process® and high pressure processed as a function of NaCl, KCl, K-lactate and water contents", *Innov. Food Sci. Emerg. Technol.,* vol. 13, pp. 69-74, 2012.
[http://dx.doi.org/10.1016/j.ifset.2011.09.005]

[45] K. Stollewerk, A. Jofré, J. Comaposada, J. Arnau, and M. Garriga, "Food safety and microbiological quality aspects of QDS process® andhigh pressure treatment of fermented fish sausages", *Food Control,* vol. 38, no. 1, pp. 130-135, 2014.
[http://dx.doi.org/10.1016/j.foodcont.2013.10.009]

[46] J. Comaposada, Impact of the dryness level on the quality of fermented sausages produced by means of Quick-Dry-Slice process (QDS process ®) technology., *Fleischwirtschaft Int.,* vol. 25, no. 2, pp. 215-224, 2006.

[47] G. Ferrini, J. Arnau, M.D. Guàrdia, and J. Comaposada, "The effect of thermal processing condition on the physicochemical and sensory characteristics of fermented sausages dried by Quick-Dry-Slice process®", *Meat Sci.,* vol. 96, pp. 688-694, 2014.
[http://dx.doi.org/10.1016/j.meatsci.2013.10.005] [PMID: 24200559]

CHAPTER 3

Meat Emulsion Technology

Abstract: The technology of the production of dry-fermented products is old and even in current days these meat products are produced, appreciated and consumed worldwide. The fermentation and drying steps are responsible for chemical and biochemical transformations that ensure the final characteristics of the product were reached. One of the disadvantages in dry-fermented product processing is the time required for all transformations that characterize the product and meets the required standard of safety and quality. Thus, the use of emerging technologies allows the improvement in the production process considering the use of new approaches, always ensuring food safety and quality. In this chapter, the ultrasound, pulsed UV light and QDS process® are discussed. The approach includes the advantages and disadvantages of each technology, the report of researchers describing the use of mentioned technologies and the most probable mechanisms associated with the effects.

Keywords: Fermented Sausages, Ham, Pulsed UV Light, Quick-Dry-Slice Process®, Salami, Starter Culture, Ultrasound.

INTRODUCTION

Emulsified meat products are widely consumed all over the world and are of great economic importance to the modern meat industry [1, 2]. Although meat batters have been a traditional age-long prepared product, the scientific principles and expertise are significantly important for commercial products [2]. Cooked sausages processed products such as mortadellas (bologna) and frankfurters (wiener, hot dog or franks) are the main meat products obtained using emulsification processes considered the most popular ready-to-eat (RTE) cooked meat products and convenient for modern lifestyles. Nowadays, emulsioned products of several species, especially beef, pork, poultry, turkey, and fish are to be found.

A meat emulsion is a mixture in which finely divided meat constituents disperse analogously to an oil-in-water emulsion. The discontinuous phase is fat, and the continuous phase is an aqueous salt and protein solution, with suspended insoluble proteins, portions of muscle fibers in the sarcolemma, and remnants of connective tissue. The main emulsifying agents are myofibrillar proteins due to

Daneysa L. Kalschne, Marinês P. Corso & Cristiane Canan

their higher solubility (saline-solution soluble) [3]. However, the use of stromal proteins, especially collagen and sarcoplasmic proteins, improved the protein concentration and emulsion stability [4, 5].

Meat emulsification is mainly affected by factors such as chopping parameters (temperature, speed, and time) and raw meat materials used (raw meat type, meat pH, type and amount of fat, non-meat ingredients, emulsifiers, and hydrophilic gums) [2, 3, 6]. Therefore, methods allowing accurate meat emulsification process control or improving stability ought to be considered. For this purpose, the ultrasound method will be addressed.

The ultrasound method has been shown to provide fat reduction for emulsified products [7, 8]. This is a quite relevant factor, considering the high-fat content in bologna sausages and frankfurters, which makes it difficult to replace animal fat and the need for the meat industry to meet consumer demand by offering healthier products. The multiple emulsion method (W1/O/W2) has also been shown to improve the technological, physicochemical, and sensory properties of emulsified products made with total or partial replacement of animal fat for non-meat fat [9, 10].

Another point that deserves special attention is that cooked frankfurters and bologna sausages packaging is typically re-exposed to the processing environment processing, leading to potential microorganism contamination. Moreover, to meet consumer demand for ready-to-eat (RTE) products, the meat industry has been providing emulsified meat products, such as bologna sausage in small portions (thin slices) from processed blocks, which could cause microbiological contamination after pasteurization [11 - 13]. Therefore, additional processes are required to ensure the safety of the product. The use of emergent methods such as ohmic heating and pulsed light as well as the use of additives [14] have been studied. The ohmic heating has been identified as a promising cooking process for its use in meat industries due to energy saving and shorter processing time.

Ultrasound, double emulsion, pulsed UV light, and ohmic heating methods are discussed as alternative processes in the preparation of emulsified meat products in this chapter.

THE TRADITIONAL EMULSIFIED PRODUCTS PREPARATION PROCESS

Frankfurters are generally made of 60% beef and 40% pork. Wieners could be prepared with 100% beef, 100% pork or 100% poultry meat, or a combination of them. Wieners may vary in size and style for different markets. Frankfurters are larger, while Vienna sausages are smaller. The Vienna sausage is also canned and could be stored at room temperature. The formulation and processing of mortadella are similar to franks, but bologna sausage casing is much larger in size than wieners [15].

Mortadella is a large emulsified cooked sausage with different denominations and composition, depending on the geographical location where it is produced. Originally from Bologna in Italy, bologna sausage is made from finely comminuted and cured pork usually added with small pork fat cubes. It has a characteristic garlic flavor and coarse black peppers are added to it in some South American countries.

Other meats such as buffalo, lamb, mutton [16], and fish might also be used in similar products. It is also noteworthy that the use of mechanically separated meat (especially poultry) is law-permitted in some countries and is common in emulsified products [17], also muscle nonskeletal meat (liver, tongue, heart, *etc.*), tendons, and skin are permitted for similar low-cost products. As for the non-meat ingredients, the main one is iced-water for keeping the batter at a low temperature, as well as salt, nitrites, antioxidants, sugars, stabilizers (phosphates and other emulsifiers) and/or water-binders such as soy protein, caseinate, and starch are used in these products. In addition to flavorings, seasonings and spices that are characteristic of each product and are dependent on regional traditions. Table 1 shows some possible emulsified meat product formulations. It is possible to notice the variability of the raw material and ingredients used, which makes it difficult to standardize these products and shelf-life. Therefore, the innovative technologies that will be discussed in this chapter are necessary to contribute to the quality and safety of these meat products.

Frankfurter and bologna sausage production involves the selection of raw material, grinding, seasoning, blending, emulsification, stuffing, cooking, and packaging. Processing up to the stuffing stage is quite similar for both products. According to Abdolghafour *et al.*, the process begins by choosing the raw material, which could be either fresh or frozen, but of good microbiological quality. All ingredients must be properly weighted according to the formulation.

The lean meat portion should be trimmed to less than 10% fat and connective tissue, free from tendons, cartilage and bones, for good water-binding capacity [16]. Meats vary in their ability to bind water and hold lean meat and fat together. Lean skeletal muscle tissue proteins contain the best water-retention and fat-emulsifying properties. Adding salt (15–25%) to the formulation for myofibrillar protein solubilization enables it to form a gel when heated [15].

Table 1. Examples of cooked emulsified sausage formulations.

Product	Raw Material	Ingredients	pH	Aw	Size	Reference
Hot dog* Sausages	Mechanically separated chicken, pork, turkey pork fat	Water, isolated soy protein, sodium chloride, sodium nitrite, sodium erythorbate, sodium lactate, sodium tripolyphosphate, sodium acid pyrophosphate, maltodextrin, white and black pepper, natural smoked aroma and monosodium glutamate, dyed with annatto	6.49	0.96	-	[18]
Vienna sausages*	Mechanically separated chicken beef pork	Water, salt, corn syrup, and less than 2% mustard, natural flavor, dried garlic, and sodium nitrite, canned in a chicken broth and caramel solution	5.7	-	20 mm diameter and an average 53 mm in length	[19]
Frankfurter batters	Lean pork muscle (320 g kg⁻¹) Pork back fat (380 g kg⁻¹)	Water (iced) (213 g kg⁻¹), seasoning GR-6304 (wheat flour, salt, isolated soy protein, dextrose, spices, tomato powder, yeast extract, stabilizers E450 and E451, smoke flavouring, hydrolysed vegetable protein, antioxidant E301, garlic powder, spice extracts and flavouring) (75 g kg⁻¹), Tari K2 (12 g kg⁻¹)	-	-	-	[20]
Bologna sausages	Pork meat (550 g kg⁻¹) Pork back-fat (350 g kg⁻¹)	Ice (100 g kg⁻¹), iodized NaCl (26 g kg⁻¹), powdered milk (12 g kg⁻¹), garlic (2 g kg⁻¹), curavi (mixture: NaCl, E-250, E-252 and antioxidant E-331) (3 g kg⁻¹), polyphosphates (Mixture: E-430i, E-454i and E-451i) (2 g kg⁻¹), sodium ascorbate (0.5 g kg⁻¹), BDRom Carne (commercial aromas) (1 g kg⁻¹), monosodium glutamate (1 g kg⁻¹), Carmin de Cochenille 50% (E-120) (0.1 g kg⁻¹)	6.2	0.92	60 mm diameter	[21]

* The raw material and ingredients described on the commercial products' label.

The raw meat is transferred to a bowl chopper (cutter). The chopping or emulsification process involves cutting and mixing meat and fat with other ingredients until obtaining an emulsion-like fine batter. The cutter is an industrial equipment composed of propellers and a rotating bowl homogenizer. As the sharp propellers rotate to cut the meat and mix the ingredients to promote the emulsion, the rotating bowl rotates for greater emulsion homogenization. Fig. (**1**) describes a scheme of cutter design.

Fig. (1). Scheme of cutter for fine emulsion preparation.

However, a meat emulsion is more similar to a meat batter than an emulsion [15]. The chopping process is crucial to meat emulsified products as the raw meat must meet the appropriate temperature, time, and pH level in order to ensure a dense texture, rich flavor, good flexibility, and optimum oil and water holding capacity [16]. Fig. (**2**) shows an emulsified meat product mass being prepared in a cutter in a Laboratory of Meat Industrialization by Food Technology and Food Engineering students.

For each product, the vacuum stuffing process involves inserting the prepared batter into appropriate casings. The vacuum provides better pieces formation with firmer consistency and color and taste preservation, which is due to a delay in fat oxidation. For mortadellas and frankfurters, collagen-based films may be used on both naturally cooked and smoked products, as well as in plastic films for cooked products without the need for smoke permeability. Mortadellas are enclosed in 2.75 to 5.90 in. casings. Such casing is used as a marketing piece (for whole-piece marketed products) or removed before slicing (for sliced-marketed products). Frankfurter stuffing occurs in 0.75 to 1.15 in diameter casings. After cooking and/or smoking, the film is removed to dye the annatto pieces and/or the brand packaging.

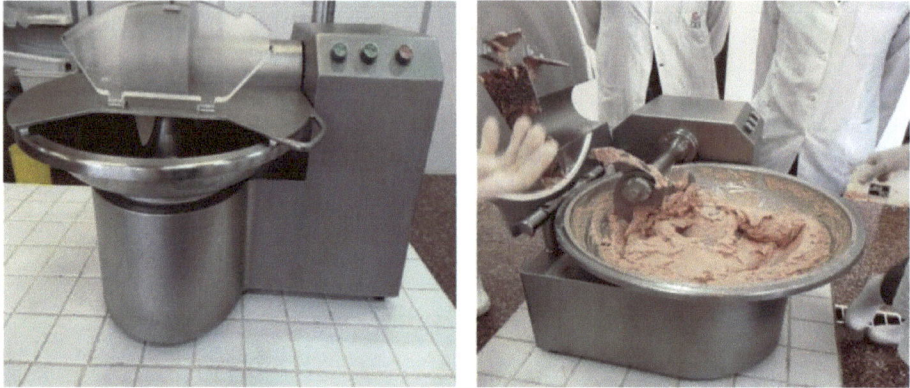

Fig. (2). Emulsified meat product mass prepared in a cutter.

For heat treatment, Pereda *et al.* related that the following effects: a) batter binding by protein coagulation, establishing a meat gel and thereby achieving the desired texture (65 to 70 ºC), in addition to developing flavor and color by curing reactions; b) meat enzymes inactivation that could cause further changes in the product; c) elimination of microorganisms vegetative cells (72 °C) are achieved in meat products cooking. Heat treatment application is usually done by immersing the pieces in hot water or steam smokehouses. The cooking period is dependent on the piece size and time to reach an internal temperature of 72 °C, thus allowing the product to be pasteurized. Overall, the cooking process is a slow one especially for larger products, causing greater energy consumption. For canned products such as Vienna sausage or presented in glass containers, the heat treatment to obtain sterile products is performed by autoclaves at 121 ºC, thus providing longer preservation, even in non-refrigerated environments [3].

Emerging methods or technologies aiming at improving emulsification, cooking, and post-pasteurization decontamination will be discussed next.

EMERGING METHODS OR TECHNOLOGIES FOR EMULSIFIED COOKED MEAT PRODUCTION

ULTRASOUND

Food emulsion preparation by acoustic cavitation using ultrasound is a widespread basic method for upgrading the sensory characteristics of the products and boosting the economics of the production process [22].

The use of ultrasound in the food industry is divided into low- and high-intensity ultrasound. Low-intensity ultrasound uses power levels (typically <1 Wcm^{-2}) small enough so the ultrasonic waves do not cause any physical or chemical disruption on the material. On the other hand, power levels used in high-intensity ultrasound (typically ranging from 10 to 1000 Wcm^{-2}) are capable to cause physical disruption of the material to which they are applied or foster certain chemical reactions (*e.g.* oxidation) [23].

The use of the ultrasound method in emulsified meat preparation technology has been mainly linked to emulsion stability. A summary of ultrasound method conditions, the product applied, and the main results obtained are seen in Table **2**.

Table 2. Ultrasound method application researches on emulsified meat preparation.

Product	Process conditions	Main results	Reference
PSE-like meat	20 kHz, 450 W, 60% amplitude, for 0, 3, and 6 min	The 6-min treatment increased gel strength, viscosity-elasticity, and WHC, giving them properties similar to normal meat gels.	[24]
Chicken actomiosin solution	20 kHz ultrasound probe; 100, 150 and 200 W (1.15, 2.36 and 11.43 Wcm^{-2}), respectively for 2 s and off-time 4 s was given for total 20 min	The conformation changes of treated chicken actomiosin showed a reduction in α-helix, increase protein hydrophobicity and reactive SH groups. Solubility and emulsion property increased significantly at 150 W. However, the emulsion stability index had a gradual decrease with increased ultrasound power	[25]
Beef myofibrillar proteins	20 kHz, 100, and 300 W for 10, 20, and 30 min	Increased time and ultrasound power (30 min at 300 W) led to enhancing pH, water holding capacity, gel strength, and emulsifying properties enhancement.	[26]
Meat emulsion	25 kHz, 60% amplitude, 19.25 W L^{-1}, 1000 W nominal power, and 154 W acoustic power (ultrasonic bath) for 5.5 min. Operating modes: degas, normal, and sweep	Cavitation distribution in a normal operating mode significantly favored a higher yield (88.7%) and emulsion stability, with good cohesiveness (0.76), hardness (26.9 N), and chewiness (26.1 N) values, not increasing lipid and protein oxidation.	[27]
Bologna sausage with reduced phosphate	25 kHz frequency, 230 W acoustic power, 33 WL^{-1} volumetric power, 60% amplitude for 0, 9, and 18 min	The use of ultrasound for 18 min improved the technological quality and did not increase the lipid oxidation, reducing most of the sensory defects caused by a 50% phosphate level reduction in bologna sausage.	[1]
Soybean oil emulsions replacing animal fat in meat emulsion	20 kHz ultrasonic probe, 450 W for 0, 3, 6, 9, and 12 min (on- and off-time pulse durations of 4 and 2 s, respectively)	Oil emulsion prepared from a 6-min ultrasound treatment showed smaller oil droplets and a more ordered, higher density microstructure. The emulsion-type meat products resulted in a more viscoelastic network and improved TPA parameters (hardness, chewiness, *etc.*).	[7]

(Table 2) cont.....

Product	Process conditions	Main results	Reference
Soybean oil emulsions replacing animal fat in frankfurters	20 kHz using a 1/2 "-diameter tip operating at 60 μm amplitude and 30-40 W power 3 input for 30 min	Reformulated frankfurters were less red and lighter presenting hardness similar to traditional ones, but sensory scores showed lower acceptability than the control.	[8]
Chicken myofibrillar protein	20 kHz frequency, 450 W power, 30 Wcm^{-2} intensity for 0, 3, and 6 min	High-intensity ultrasound (6 min) significantly improved the emulsifying properties of chicken myofibrillar protein enhancing the rheological properties and storage ability of MP-stabilized emulsion.	[28]

Based on that, some work has been done using high-intensity ultrasound to improve the technological properties of the raw material for emulsified meat preparation. Zou *et al.* observed that ultrasound treatment at 100–150 W might be useful to modify chicken actominosin, thereby improving its functional properties. The unfolding, stretching, and dissociation of chicken actomiosin, as well as subsequent exposure of hydrophobic residues by ultrasound treatment, indicated that emulsifying property was enhanced [25]. Li *et al.* studied the effects of high-intensity ultrasound on the emulsifying properties of the chicken myofibrillar protein, the rheological properties and the emulsion stability. Ultrasound increased the unfolding of myofibrillar proteins increasing the reactive sulfhydryl content, surface hydrophobicity, and intrinsic fluorescence intensity. Ultrasound significantly reduced the α-helical content and increased the β-sheet, β-turn, and random coil contents. These structural changes increased interfacial proteins around oil droplets, enhancing the emulsifying properties [28]. Similar effects were also observed for beef myofibrillar protein [26].

High-intensity ultrasound may also be used to modify the functionality of pale, soft, exudative-like (PSE-like) low-quality raw meat materials. Li *et al.* applied high-intensity ultrasound in PSE-like chicken breast meat and the results showed emulsification and gelling enhancements offering potential economic benefits for the poultry industry [24].

According to Cichosky *et al.*, the use of ultrasound proved to be an effective tool in emulsions preparation. However, the cavitation phenomenon increases temperature, which may affect emulsion stability and texture. The authors verified that the wave propagation mode in the ultrasound bath exerts an influence on the temperature increase of meat emulsions. Cavitation distribution in a normal operating mode (providing a frequency stabilization, fostering a continuous flow and better in-bath wave distribution) favored a higher yield and emulsion stability, not increasing lipid and protein oxidation [27].

According to Krasulya *et al.*, two approaches to the ultrasonic preparation of emulsions must be considered in addition to raw material ultrasonic rattling: a) direct sonication of the system generating emulsion by acoustic cavitation; b) cavitation activation of a continuous medium (*e.g.* brines), as a result of which water acquires unique properties related to its structural changes. The emulsion prepared using cavitation-treated brine impairs the sausage structure formation. The impact caused by the intensity creates emulsion drops smaller than 15 μm. The product is obtained with a smooth elastic texture and a pronounced taste, making it more attractive to the customer [22].

In addition to emulsion stability enhancement, other applications have also been addressed. Cichosky *et al.* suggested the use of ultrasound (US) to reduce temperature and accelerate packaged hot dog sausages' pasteurization process. They were pasteurized in an ultrasonic bath (US) (25 kHz, 200 W) for 10.53 min at 74 °C and also without US, as well as conventional pasteurization, with water bath at 82 °C, for 16 min to reach a 73 °C internal sausage temperature. US treatment inhibited the growth of psychotropic and lactic bacteria, reduced lipid oxidation fostering slight pH and texture modifications during storage, enhancing pasteurization [18].

Ultimately, another approach takes into account the high fat, high sodium, and chemical additive contents found on emulsified meat products such as bologna sausage and sausages; thus, they were targeted for food reformulation in order to enhance its nutritional quality and consumers' health. Ultrasound could be an alternative for developing healthier emulsified meat products [1]. Pinton *et al.* suggested the possibility of producing bologna sausages with an up to 50% phosphate content reduction by sonicating the pieces in an ultrasonic bath maintaining their technological, oxidative, and sensory qualities [1]. Zhao *et al.* demonstrated the potential of high-intensity ultrasound on pre-formed soybean oil emulsions preparation to replace animal fat and reduce the saturated fatty acid content in meat products. Rheological tests in emulsions elaborated with chicken breast myofibrillar protein showed that ultrasound-treated samples formed a more viscoelastic gel than the control one. Water- and fat-binding and textural properties were also significantly improved by pulsed ultrasound [7]. Paglarini *et al.*, aimed at replacing animal fat in frankfurters by using emulsion gels prepared with soybean oil, sonicated and non-sonicated soy protein isolate dispersions, carrageenan, and inulin. They noted that restructured soybean oil and functional ingredients might be suitable in texture terms to replace fat in meat products, which are greatly affected when the oils are added in their natural form. However, animal fat is hard to be replaced and acceptance tests showed that sensory properties on samples need to be improved. The authors suggested a partial replacement using pork back fat [8].

Such approaches demonstrate the potential of ultrasound methods on improving the meat product quality and making better use of raw materials. Some limitations suggest that further studies are required in order to optimize results and provide their application in the meat industry.

DOUBLE EMULSIONS (DEs)

Various studies showed that in recent years, there has been a growing interest in the reduction and enhancement of meat products fat contents in order to meet consumer demands. However, there has been great difficulty in replacing animal fat due to changes in the sensory and physicochemical properties of products. In this context, DEs have been proposed as a novelty for lipid profile improvement on animal fat replacement usually used in foods with high saturated fat content [10].

Water-in-oil-in-water (W1/O/W2) multiple emulsions are multi-compar--mentalized systems where water-in-oil (W1/O) emulsion droplets are dispersed within an outer continuous aqueous phase (W2) [29].

The double emulsifying method (DEs) is, among others, one of the most used as it obtains stable and well-defined systems with reproducible particle sizes, *i.e.* it is used as a pre-emulsion method to stabilize the incorporated non-meat fats. These oil-in-water pre-emulsions are stabilized by non-meat proteins (such as sodium caseinate or soy protein isolate) and used as fat ingredients [30 - 33].

Cofrades *et al.* studied the double-emulsion method in order to replace animal fat in a meat emulsion system, the method of which has been applied to dairy products fat reduction and encapsulation of bioactive compounds. For that, W1/O/W2 emulsions prepared with olive oil as lipid phase, polyglycerol ester of polyricinoleic acid as a lipophilic emulsifier, sodium caseinate, and whey protein concentrate as hydrophilic emulsifiers presented good stability for longer than that required by the meat industry standards. Therefore, they affected the water and fat binding properties, texture, and color of gel/emulsion meat systems when used as animal fat (pork back fat) replacers [31] (Fig. **3**).

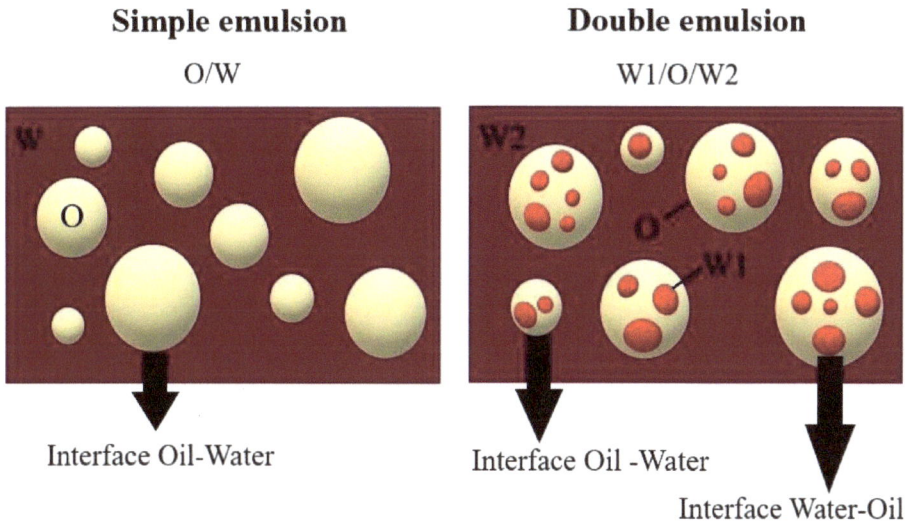

Figure (3). Emulsified meat product mass prepared in a cutter (W: water; O: oil).

Numerous studies with promising results were performed using pre-emulsified vegetable oils as a total or partial replacement for pork fat. Freire *et al*. achieved over 60% fat reduction in frankfurters with the replacement of pork fat by two different W/O/W emulsions containing perilla oil and pork back fat [32]. Kang *et al*. obtained a lower fat and energy contents replaced pork back fat by pre-emulsified sesame oil, resulting in greater pork batter texture [34]. Eisinaite *et al*. used double emulsions to replace 7 and 11% animal fat and to enhance the meat product's color. The coarse emulsion contained native beetroot juice as the inner water phase, sunflower oil as the oil phase, and whey protein isolate as the outer water phase. The results showed that double emulsion technology could be used to achieve two functionalities in meat products: caloric load reduction and color retention [9]. More recently, Robert el al. also showed that the use double emulsions containing healthy oil blend (olive, linseed, and fish oils (70:20:10 w/w/w)) as lipid phase and an olive leaves extract encapsulated as the internal aqueous phase could be incorporated as fat replacers in meat systems, enhancing both the lipid profile and the oxidative stability [10]. Kiokias *et al*. stressed the scarcity of research on multiple emulsions oxidative stability. Given its relevance, oxidative stability must be verified before applying multiple emulsions, especially when using a combination of highly unsaturated oils as a lipid phase [33].

OHMIC HEATING

Conventional heating methods are based on conduction and convection heat transfer modes and are the most widely used methods in the food industry. Thermal processing is applied to food for various reasons, such as palatability enhancement and shelf life extension [35]. For meat emulsion products, the cooking process involves either steam cooking or water immersion methods. The low heat penetration rate to the thermal center, especially for large-sized products, is a major issue for meat cooking, which could lead to long cooking periods causing a more severe heat exposure to the outer layers than to the inner parts. This could potentially impair the quality of the outer parts of such products [36]. Electro-heating methods such as ohmic heating were studied and seen as promising cooking processes for use in meat industries saving on both energy consumption and processing time.

For the ohmic heating process, an alternating electric current passes through a conductor and the flowing charges increase temperature according to Joule's law [35] *i.e.*, the method uses the meat products resistance to convert the electric energy into heat. The heat generation rate depends on the voltage gradient applied and the electrical conductivity of the meat product [37]. Unlike conventional heating methods that rely on heat transfer from a heating surface, ohmic heating generates heat volumetrically inside the material and increases the temperature at a higher rate [35]. Thus, the advantages of ohmic cooking over conventional heating include shorter processing time, higher yield, and less power consumption while still maintaining the meat products' color and nutritional value [37].

Piette *et al.* cooked 1 kg basic bologna emulsion portions, either in a smokehouse (180-min cycle; to 70 °C at core) or by ohmic heating (64 to 103 V; from 3.9 °C min^{-1} to 10.3 °C min^{-1}; from 70 °C to 80 °C). The research showed that ohmic-cooked sausages were similar to smokehouse-cooked ones regarding color, texture, pH, drip, redox potential, and rancidity, except for texture. The results showed the ohmic-cooked product was softer, less cohesive, and less resilient than the smokehouse-cooked one. The authors also observed that a soft textured sausage is more appealing to some consumers as the measured values fell within the range typical of commercial sausages [38].

In a similar study, Shirsat *et al.* cooked frankfurters in a 3.5 kW batch ohmic heater (230 V, 50 Hz, 3, 5, and 7 Vcm^{-1}) and steam (80 °C to give product endpoint temperatures higher than 73 °C for 2 min). The ohmic heating yielded a product with a lower springiness suggesting lower elasticity as the reported TPA in ohmic-cooked samples and differed from steam-cooked ones in both a* and hue angle values (especially at 5–7 Vcm^{-1}) (p < 0.05). However, tasters were not able to distinguish samples in the sensory analysis [20].

The results obtained by Piette *et al.* suggested that the quality of ohmic-cooked sausages would not be much influenced by changes in cooking cycles required for processing convenience (*i.e.* time) or to ensure microbiological safety. The extent of bacterial inactivation during ohmic and smokehouse cooking was calculated using *Enterococcus faecalis*, which is commonly used as a reference for testing bacterial inactivation in cooked meats [38].

More recently, ohmic heating was studied in emulsified meat products focusing on post-pasteurization decontamination. In the work by Inmanee *et al.*, ohmic heating of vacuum-packed sausages was suggested for effective post-decontamination packaging. Ohmic heating at 75 °C for 30 s was investigated and compared with two conventional heating ranges at 75 °C for 30 s and 75 °C for 2 min regarding *L. monocytogenes* inactivation. The authors also investigated the effects of the conditions on product properties and sensory attributes. The results showed that the pathological bactericidal power on *L. monocytogenes* using ohmic heating was higher compared to conventional heating with a greater total cooking time reduction. Moreover, no change was observed in either pH, lipid oxidation, cooking loss or water holding capacity apart from a few alterations to texture and color under ohmic heating. Therefore, ohmic heating showed potential in the post-pasteurization decontamination for sausage production, with a minimal negative effect on quality [39].

PULSED LIGHT

Ready-to-eat processed meats, especially sliced ones, represent a concern for the meat industry since works showed the risk of cross-contamination by microorganisms such as *L. monocytogenes*, *L. innocua* and *Salmonella* in this type of product is a hazard to consumers health with severely harmful effects [11 - 13]. For this, methods that could be applied post-cooking and post-packaging for decontamination without changing the sensory quality of the product has been investigated.

Non-thermal disinfection methods have been gaining increasing interest in this field. The use of broad spectral range (200–1100 nm) intense light pulses, comprising of ultraviolet, visible, and near-infrared irradiation is currently seen as a potentially suitable technology for reducing microbial loads on different food surfaces. Pulsed light causes photochemical changes on nucleic acids by UV-C in addition to photothermal and photophysical damage to cells by membrane disruption and water vaporization, creating a decontamination effect [40 - 43]. Since light penetration occurs only at a superficial level, it would not be a problem for RTE foods as post-heat treatment contamination takes place on the surface [14].

FDA-approved as a non-thermal treatment since 1996 [44], pulsed light has already been developed for commercial-scale food packaging, but with little evidence for large scale food treatments. Emerging applications include meat and fish products and associated packaging decontamination [45]. The use of pulsed light for emulsified meat products has been highlighted with promising results.

Uesugi and Moraru investigated the antilisterial capability by combining pulsed light with the natural antimicrobial nisin. The treatments were performed using commercial samples of Vienna sausages individually centered on an adjustable stainless steel shelf in the pulsed light unit 50.8 mm beneath the xenon lamp and treated with a varied number of pulses at three pulses per second frequency, and a 360 μs pulse width. PL reduced *L. innocua* (used as an indicator for the pathogenic *L. monocytogenes*) by 1.37 log CFU after exposure to 9.4 Jcm^{-2}. A total reduction of 4.03 log CFU was recorded after the combined treatment of nisin and PL for 48 h at 4° C. The treatment resulted in no significant microbial growth during a 48-dayrefrigerated storage period. The author also mentioned applying it through UV-transparent packaging material, which would represent a quality step in ensuring the safety of RTE food products [19]. Such possibility was also observed by Keklik *et al*. Pulsed UV-light was effective on *L. monocytogenes* on the surfaces of unpackaged and vacuum-packaged (with polypropylene) chicken frankfurters. The optimum treatment conditions for both were found to be at 8 cm for 60 s, which resulted in about 1.6 and 1.5 log CFU.cm^{-2} reductions (both at approximately 97%) on unpackaged and vacuum-packaged samples, respectively. 1.9 CFU.cm^{-2} maximum log reduction (approximately 99%) was observed after unpackaged and vacuum-packaged samples treatment at 5 cm for 60 s, which caused visual and quality changes to the samples [46].

Pulsed light was also tested in vacuum-packaged bologna sausage slices superficially inoculated with *Listeria monocytogenes* and treated with 0.7, 2.1, 4.2, and 8.4 J.cm^{-2}. PL treatment at 8.4 J.cm^{-2} reduced *L. monocytogenes* by 1.11 cfu.cm^{-2} in bologna sausage. Therefore treatments above 2.1 J.cm^{-2} negatively influence the sensory properties of bologna sausage [14]. The results indicate the need for further studies.

The type of film should also be considered for packaged product treatment as the UV-C light transparency of the films could, for instance, be different and therefore changing the potential for decontamination depending on the film used on packaging the meat product [47].

FINAL CONSIDERATIONS

Emerging methods for cooked emulsified meat product elaboration, such as frankfurters and mortadellas, addressed in this chapter have proven their potential for industrial applications. The use of ultrasound and double emulsion method is promising for emulsion stability, allowing the use of low-quality raw materials such as PSE meat, and saturated pork fat and chemical additives reducers. Other methods such as ohmic heating might be an alternative to lower energy consumption and cooking time, especially in large-sized sausages such as bolognas. The use of pulsed light for frankfurters and sliced mortadellas decontamination is another relevant advancement, making these ready-to-eat products safer for consumers. However, studies in order to obtain optimal process conditions aiming at the efficiency with minimal changes in the physical, chemical and sensory characteristics of sausages, as well as equipment implementation for industrial-scale should be encouraged.

REFERENCES

[1]   M.B. Pinton, L.P. Correa, M.M.X. Facchi, R.T. Heck, Y.S.V. Leães, A.J. Cichoski, J.M. Lorenzo, M. Dos Santos, M.A.R. Pollonio, and P.C.B. Campagnol, "Ultrasound: A new approach to reduce phosphate content of meat emulsions", *Meat Sci.,* vol. 152, pp. 88-95, 2019.
[http://dx.doi.org/10.1016/j.meatsci.2019.02.010] [PMID: 30836267]

[2]   D. Santhi, A. Kalaikannan, and S. Sureshkumar, "Factors influencing meat emulsion properties and product texture: A review", *Crit. Rev. Food Sci. Nutr.,* vol. 57, no. 10, pp. 2021-2027, 2017.
[http://dx.doi.org/10.1080/10408398.2013.858027] [PMID: 25836950]

[3]   M.D.S.C.J.A.O. Pereda, M.I.C. Rodriguez, L.F. Álvarez, M.L.G. Sanz, G.D.G.F. Minguillón, and L.H. Pereales, *Tecnologia de Alimentos. Produtos de Origem Animal.* Artmed: Porto Alegre, 2005.

[4]   R.C. Santana, F.A. Perrechil, A.C.K. Sato, and R.L. Cunha, "Emulsifying properties of collagen fibers: Effect of pH, protein concentration and homogenization pressure", *Food Hydrocoll.,* vol. 25, no. 4, pp. 604-612, 2011.
[http://dx.doi.org/10.1016/j.foodhyd.2010.07.018]

[5]   O.A. Young, S.X. Zhang, M.M. Farouk, and C. Podmore, "Effects of pH adjustment with phosphates on attributes and functionalities of normal and high pH beef", *Meat Sci.,* vol. 70, no. 1, pp. 133-139, 2005.
[http://dx.doi.org/10.1016/j.meatsci.2004.12.018] [PMID: 22063289]

[6]   C.U.I. Yanfei, Z. Gaiming, W. Yufen, X.I.E. Hua, L.I. Miaoyun, L.I.U Yanxia, and X.U. Xiong, "Advance on Effective Chopping Emulsification Technology in Meat Processing", Online Available: file:///D:/NaoApagar/Downloads/%E9%B1%BC%E7%B3%9C%E5%87%9D%E8%83%B6%E6%80 %A7%E8%83%BD%E7%A0%94%E7%A9%B6%E8%BF%9B%E5%B1%95.pdf [Accessed Jul. 22, 2020].

[7]   Y.Y. Zhao, "Effect of pre-emulsification of plant lipid treated by pulsed ultrasound on the functional properties of chicken breast myofibrillar protein composite gel", *Food Res. Int.,* vol. 58, pp. 98-104, 2014.
[http://dx.doi.org/10.1016/j.foodres.2014.01.024]

[8]   C. de S. Paglarini, S. Martini, and M.A.R. Pollonio, "Using emulsion gels made with sonicated soy protein isolate dispersions to replace fat in frankfurters", *Lwt,* vol. 99, pp. 453-459, 2018.
[http://dx.doi.org/10.1016/j.lwt.2018.10.005]

[9] V. Eisinaite, D. Juraite, K. Schroën, and D. Leskauskaite, "Food-grade double emulsions as effective fat replacers in meat systems", *J. Food Eng.,* vol. 213, pp. 54-59, 2017.
[http://dx.doi.org/10.1016/j.jfoodeng.2017.05.022]

[10] P. Robert, M. Zamorano, E. González, A. Silva-Weiss, S. Cofrades, and B. Giménez, "Double emulsions with olive leaves extract as fat replacers in meat systems with high oxidative stability", *Food Res. Int,* vol. 120, pp. 904-912, 2019.
[http://dx.doi.org/10.1016/j.foodres.2018.12.014]

[11] M.E. Berrang, R.J. Meinersmann, J.K. Northcutt, and D.P. Smith, "Molecular characterization of Listeria monocytogenes isolated from a poultry further processing facility and from fully cooked product", *J. Food Prot.,* vol. 65, no. 10, pp. 1574-1579, 2002.
[http://dx.doi.org/10.4315/0362-028X-65.10.1574] [PMID: 12380741]

[12] R.Y. Murphy, L.K. Duncan, B.L. Beard, and K.H. Driscoll, "D and z values of Salmonella, Listeria innocua, and Listeria monocytogenes in fully cooked poultry products", *J. Food Sci.,* vol. 68, no. 4, pp. 1443-1447, 2003.
[http://dx.doi.org/10.1111/j.1365-2621.2003.tb09664.x]

[13] K.L. Vorst, E.C.D. Todd, and E.T. Ryser, "Transfer of Listeria monocytogenes during slicing of turkey breast, bologna, and salami with simulated kitchen knives", *J. Food Prot.,* vol. 69, no. 12, pp. 2939-2946, 2006.
[http://dx.doi.org/10.4315/0362-028X-69.12.2939] [PMID: 17186662]

[14] E. Hierro, E. Barroso, L. De La Hoz, J.A. Ordóñez, S. Manzano, and M. Fernández, "Efficacy of pulsed light for shelf-life extension and inactivation of Listeria monocytogenes on ready-to-eat cooked meat products", *Innov. Food Sci. Emerg. Technol.,* vol. 12, no. 3, pp. 275-281, 2011.
[http://dx.doi.org/10.1016/j.ifset.2011.04.006]

[15] S.M. Lonergan, D.N. Marple, and D.G. Topel, "Sausage processing and production", In: *The Science of Animal Growth and Meat Technology* 2nd ed. Academic Press, 2019, pp. 229-253.
[http://dx.doi.org/10.1016/C2017-0-02648-4]

[16] B. Abdolghafour, and A. Saghir, "Development in sausage production and practices-A review Sausage Production : Ingredients and Raw", *J. Mater. Sci. Technol.,* vol. 2, no. 3, pp. 40-50, 2014.

[17] F.G. Daros, M.L. Masson, and S.C. Amico, "The influence of the addition of mechanically deboned poultry meat on the rheological properties of sausage", *J. Food Eng.,* vol. 68, no. 2, pp. 185-189, 2005.
[http://dx.doi.org/10.1016/j.jfoodeng.2004.05.030]

[18] A.J. Cichoski, "Ultrasound-assisted post-packaging pasteurization of sausages", *Innov. Food Sci. Emerg. Technol.,* vol. 30, pp. 132-137, 2015.
[http://dx.doi.org/10.1016/j.ifset.2015.04.011]

[19] A.R. Uesugi, and C.I. Moraru, "Reduction of Listeria on ready-to-eat sausages after exposure to a combination of pulsed light and nisin", *J. Food Prot.,* vol. 72, no. 2, pp. 347-353, 2009.
[http://dx.doi.org/10.4315/0362-028X-72.2.347] [PMID: 19350979]

[20] N. Shirsat, N.P. Brunton, J.G. Lyng, B. McKenna, and A. Scannell, "Texture, colour and sensory evaluation of a conventionally and ohmically cooked meat emulsion batter", *J. Sci. Food Agric.,* vol. 84, no. 14, pp. 1861-1870, 2004.
[http://dx.doi.org/10.1002/jsfa.1869]

[21] I. Berasategi, M. García-Íñiguez de Ciriano, Í. Navarro-Blasco, M.I. Calvo, R.Y. Cavero, I. Astiasarán, and D. Ansorena, "Reduced-fat bologna sausages with improved lipid fraction", *J. Sci. Food Agric.,* vol. 94, no. 4, pp. 744-751, 2014.
[http://dx.doi.org/10.1002/jsfa.6409] [PMID: 24105447]

[22] O. Krasulya, V. Bogush, V. Trishina, I. Potoroko, S. Khmelev, P. Sivashanmugam, and S. Anandan, "Impact of acoustic cavitation on food emulsions", *Ultrason. Sonochem.,* vol. 30, pp. 98-102, 2016.
[http://dx.doi.org/10.1016/j.ultsonch.2015.11.013] [PMID: 26603612]

[23]   D.J. McClements, "Advances in the application of ultrasound in food analysis and processing", *Trends Food Sci. Technol.,* vol. 6, no. 9, pp. 293-299, 1995.
[http://dx.doi.org/10.1016/S0924-2244(00)89139-6]

[24]   K. Li, Z.L. Kang, Y.Y. Zhao, X.L. Xu, and G.H. Zhou, "Use of high-intensity ultrasound to improve functional properties of batter suspensions prepared from pse-like chicken breast meat", *Food Bioprocess Technol.,* vol. 7, no. 12, pp. 3466-3477, 2014.
[http://dx.doi.org/10.1007/s11947-014-1358-y]

[25]   Y. Zou, P. Xu, H. Wu, M. Zhang, Z. Sun, C. Sun, D. Wang, J. Cao, and W. Xu, "Effects of different ultrasound power on physicochemical property and functional performance of chicken actomyosin", *Int. J. Biol. Macromol.,* vol. 113, pp. 640-647, 2018.
[http://dx.doi.org/10.1016/j.ijbiomac.2018.02.039] [PMID: 29428384]

[26]   A. Amiri, P. Sharifian, and N. Soltanizadeh, "Application of ultrasound treatment for improving the physicochemical, functional and rheological properties of myofibrillar proteins", *Int. J. Biol. Macromol.,* vol. 111, pp. 139-147, 2018.
[http://dx.doi.org/10.1016/j.ijbiomac.2017.12.167] [PMID: 29307807]

[27]   A.J. Cichoski, M.S. Silva, Y.S.V. Leães, C.C.B. Brasil, C.R. Menezes, J.S. Barin, R. Wagner, and P.C.B. Campagnol, "Ultrasound: A promising technology to improve the technological quality of meat emulsions", *Meat Sci,* vol. 148, pp. 150-155, 2019.
[http://dx.doi.org/10.1016/j.meatsci.2018.10.009]

[28]   K. Li, L. Fu, Y.Y. Zhao, S.W. Xue, P. Wang, X.L. Xu, and Y.H. Bai, "Use of high-intensity ultrasound to improve emulsifying properties of chicken myofibrillar protein and enhance the rheological properties and stability of the emulsion", *Food Hydrocoll,* vol. 98, p. 105275, 2020.
[http://dx.doi.org/10.1016/j.foodhyd.2019.105275]

[29]   D.J. McClements, "Advances in fabrication of emulsions with enhanced functionality using structural design principles", *Curr. Opin. Colloid Interface Sci.,* vol. 17, no. 5, pp. 235-245, 2012.
[http://dx.doi.org/10.1016/j.cocis.2012.06.002]

[30]   F. Jiménez-Colmenero, "Potential applications of multiple emulsions in the development of healthy and functional foods", *Food Res. Int.,* vol. 52, no. 1, pp. 64-74, 2013.
[http://dx.doi.org/10.1016/j.foodres.2013.02.040]

[31]   S. Cofrades, I. Antoniou, M.T. Solas, A.M. Herrero, and F. Jiménez-Colmenero, "Preparation and impact of multiple (water-in-oil-in-water) emulsions in meat systems", *Food Chem.,* vol. 141, no. 1, pp. 338-346, 2013.
[http://dx.doi.org/10.1016/j.foodchem.2013.02.097] [PMID: 23768366]

[32]   M. Freire, R. Bou, S. Cofrades, M.T. Solas, and F. Jiménez-Colmenero, "Double emulsions to improve frankfurter lipid content: impact of perilla oil and pork backfat", *J. Sci. Food Agric.,* vol. 96, no. 3, pp. 900-908, 2016.
[http://dx.doi.org/10.1002/jsfa.7163] [PMID: 25752293]

[33]   S. Kiokias, and T. Varzakas, "Innovative applications of food-related emulsions", *Crit. Rev. Food Sci. Nutr.,* vol. 57, no. 15, pp. 3165-3172, 2017.
[http://dx.doi.org/10.1080/10408398.2015.1130017] [PMID: 26853234]

[34]   Z.L. Kang, D.Y. Zhu, B. Li, H.J. Ma, and Z.J. Song, "Effect of pre-emulsified sesame oil on physical-chemical and rheological properties of pork batters", *Food Sci. Technol.,* vol. 37, no. 4, pp. 620-626, 2017.
[http://dx.doi.org/10.1590/1678-457x.28116]

[35]   M. Gavahian, B.K. Tiwari, Y.H. Chu, Y. Ting, and A. Farahnaky, "Food texture as affected by ohmic heating: Mechanisms involved, recent findings, benefits, and limitations", *Trends Food Sci. Technol.,* vol. 86, pp. 328-339, 2019.
[http://dx.doi.org/10.1016/j.tifs.2019.02.022]

[36]   B.M. McKenna, J. Lyng, N. Brunton, and N. Shirsat, "Advances in radio frequency and ohmic heating of meats", *J. Food Eng.,* vol. 77, no. 2, pp. 215-229, 2006.
[http://dx.doi.org/10.1016/j.jfoodeng.2005.06.052]

[37]   G. Yildiz-Turp, I.Y. Sengun, P. Kendirci, and F. Icier, "Effect of ohmic treatment on quality characteristic of meat: a review", *Meat Sci.,* vol. 93, no. 3, pp. 441-448, 2013.
[http://dx.doi.org/10.1016/j.meatsci.2012.10.013] [PMID: 23273448]

[38]   M.D.G. Piette, M.L. Buteau, D. de Halleux, L. Chiu, Y. Raymond, and H.S. Ramaswamy, "Ohmic cooking of processed meats and its effects on product quality", *JFS Food Eng. Phys. Prop.,* vol. 69, no. 2, pp. 71-78, 2004.
[http://dx.doi.org/10.1111/j.1365-2621.2004.tb15512.x]

[39]   P. Inmanee, P. Kamonpatana, and T. Pirak, "Ohmic heating effects on *Listeria monocytogenes* inactivation, and chemical, physical, and sensory characteristic alterations for vacuum packaged sausage during post pasteurization", *Lwt,* vol. 108, pp. 183-189, 2019.
[http://dx.doi.org/10.1016/j.lwt.2019.03.027]

[40]   K. Takeshita, J. Shibato, T. Sameshima, S. Fukunaga, S. Isobe, K. Arihara, and M. Itoh, "Damage of yeast cells induced by pulsed light irradiation", *Int. J. Food Microbiol.,* vol. 85, no. 1-2, pp. 151-158, 2003.
[http://dx.doi.org/10.1016/S0168-1605(02)00509-3] [PMID: 12810279]

[41]   V.M. Gómez-López, F. Devlieghere, V. Bonduelle, and J. Debevere, "Factors affecting the inactivation of micro-organisms by intense light pulses", *J. Appl. Microbiol.,* vol. 99, no. 3, pp. 460-470, 2005.
[http://dx.doi.org/10.1111/j.1365-2672.2005.02641.x] [PMID: 16108787]

[42]   J. Abida, B. Rayees, and F.A. Masoodi, "Pulsed light technology: A novel method for food preservation", *Int. Food Res. J.,* vol. 21, no. 3, pp. 839-848, 2014.

[43]   B. Kramer, J. Wunderlich, and P. Muranyi, "Recent findings in pulsed light disinfection", *J. Appl. Microbiol.,* vol. 122, no. 4, pp. 830-856, 2017.
[http://dx.doi.org/10.1111/jam.13389] [PMID: 28032924]

[44]   FDA, *Code of Federal Regulations,* 3rd ed. 21CFR179.41. Title 21, 1996. [Online] Available: www.accessdata.fda.gov/scripts/cdrh/cfdocs/cfcfr/CFRSearch.cfm?fr=179.41 [Accessed Jul. 22, 2020].

[45]   N.J. Rowan, "Pulsed light as an emerging technology to cause disruption for food and adjacent industries – Quo vadis?", *Trends Food Sci. Technol.,* vol. 88, no. March, pp. 316-332, 2019.
[http://dx.doi.org/10.1016/j.tifs.2019.03.027]

[46]   N.M. Keklik, A. Demirci, and V.M. Puri, "Inactivation of *Listeria monocytogenes* on unpackaged and vacuum-packaged chicken frankfurters using pulsed UV-light", *J. Food Sci.,* vol. 74, no. 8, pp. M431-M439, 2009.
[http://dx.doi.org/10.1111/j.1750-3841.2009.01319.x] [PMID: 19799670]

[47]   A.R. Tarek, B.A. Rasco, and S.S. Sablani, "Ultraviolet-c light inactivation kinetics of *e. coli* on bologna beef packaged in plastic films", *Food Bioprocess Technol.,* vol. 8, no. 6, pp. 1267-1280, 2015.
[http://dx.doi.org/10.1007/s11947-015-1487-y]

CHAPTER 4

Meat Curing and Preservation Methods

Abstract: Salting is one of the oldest food preserving methods. New salting methods for meat products are currently being studied aiming at reducing process time and labour costs, increasing process yield, and producing safe, low-salt content foods with acceptable sensory characteristics. In this context, ultrasound, high hydrostatic pressure and microwave technologies have been highlighted. Several manuscripts have addressed this theme and the results were satisfactory, urging the need for further studies in order to improve and enable the use of such methods. This chapter will cover traditional salting methods and new ones, currently being discussed aiming at improving the quality and yield of products.

Keywords: Brine, Dry-Cured Meat, High Pressure, Meat Quality, Microwave, Ultrasound.

INTRODUCTION

Among the different meat preservation procedures, salting is one of the most ancient methods used. However, due to the development of new food preservation methods, the meat industry currently uses salt as flavouring or flavour enhancer and also as responsible for the desirable textural properties of processed meats [1, 2]. Salt plays an essential role in meat products *e.g.* in protein it activates hydration and water-binding capacity leading to a decrease in fluid loss in thermally-processed vacuum-packaged products; an increase in the binding properties of proteins to improve texture; an increase in meat batter viscosity, facilitating the incorporation of fat to form stable batters; as well as being essential to flavouring and having a bacteriostatic feature at relatively high levels [2].

The terms salting and healing or salty and cured have, for some, the same meaning. Salted meats are also called cured meats when treated with table salt and/or curing and adjuvant salts. Salting and curing are widespread methods for obtaining meat processed products and also, in the original sense of the word, to keep meat fresh for longer.

Sodium nitrite is often included in the brine and is responsible for the cured meat colour, adding flavour, and preventing spores from microorganisms such as *Clos-*

Daneysa L. Kalschne, Marinês P. Corso & Cristiane Canan

tridium botulinum [3] from developing. A widely used adjuvant is sugar that enhances the taste of meat and counteracts some of the hardening effects caused by salt. Over time, the term meat curing became a synonym to adding salt, sugar, spices, saltpetre (nitrate) or nitrite to meat for preservation and flavour enhancement.

Salt penetration in meat is related to establishing a balance between salt concentrations inside and outside the piece of meat. However, there are a number of factors that influence the speed at which equilibrium is established, as described in Fig. (**1**) [4 - 6].

TRADITIONAL SALTED PRODUCTS

Salted or cured meat products are widely consumed worldwide. Table **1** presents some of these products and the traditional methods used for their preparation and countries of origin. It is considered that this general knowledge is necessary to arouse the interest in the application of new technologies, since the technical literature has focused mainly on hams.

Table 1. Salted meat products from different countries.

Salted Product	Meat Base	Curing Methods	Origin Country	Reference
Jerked Beef	Beef	Dry curing	Brazil	[7]
Charque	Beef, goat	Dry curing	Brazil	[7]
Cecina	Horsemeat	Dry curing	Spain	[8, 9]
Biltong	Pieces of pork, beef, goat, venison, and horse	Dry curing	South Africa	[10]
Bresaola	Pieces of pork, beef, goat, venison, and horse	Dry curing	Italy	[11]
Spanish Serrano ham	Pork	Dry curing	Spain	[12]
Iberian ham	Pork	Dry curing	Spain	[12, 13]
Italian Parma ham	Pork	Dry curing	Italy	[12]
French Bayonne ham	Pork	Dry curing	France	[12]
Lacón	Pork	Dry curing	Spain	[12, 14]
Bacon	Pork	Brine curing and/or dry curing	* Great Britain	[15]

* It is not known for sure, but records show that Britain's traditional breakfast in 1560 was served with bacon and eggs [15].

CURING METHODS

Two main forms of applying curing ingredients are dry cure and pickle or brine cure. The dry cure process involves hand rubbing the dry-cure mixture (sugar, salt, nitrate or nitrite) in excess over the external part of the meat. However, most curing procedures use a salt-based brine or pickle for manufacturing the cured meat products [3].

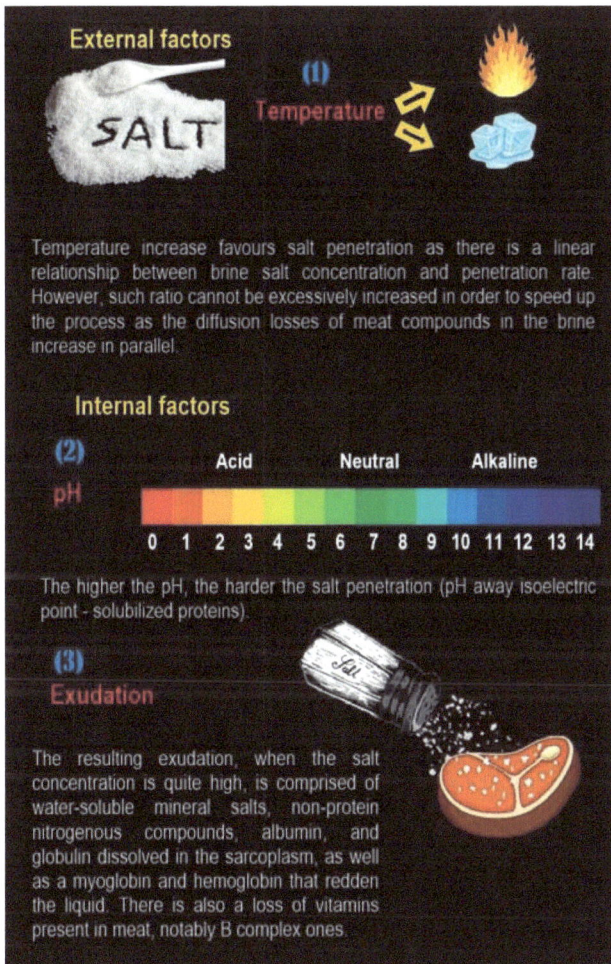

External factors

(1)
Temperature

Temperature increase favours salt penetration as there is a linear relationship between brine salt concentration and penetration rate. However, such ratio cannot be excessively increased in order to speed up the process as the diffusion losses of meat compounds in the brine increase in parallel.

Internal factors

(2)
pH

Acid   Neutral   Alkaline

0 1 2 3 4 5 6 7 8 9 10 11 12 13 14

The higher the pH, the harder the salt penetration (pH away isoelectric point - solubilized proteins).

(3)
Exudation

The resulting exudation, when the salt concentration is quite high, is comprised of water-soluble mineral salts, non-protein nitrogenous compounds, albumin, and globulin dissolved in the sarcoplasm, as well as a myoglobin and hemoglobin that redden the liquid. There is also a loss of vitamins present in meat, notably B complex ones.

Fig. (1). Factors that influence the speed and equilibrium of salted meat products.

Dry-salting might be an extremely time-consuming process and it often requires the constant rotating of meat pieces in order to standardize the product and speed up the process. No water is added, so the curing agents are solubilized in the endogenous moisture of the muscle tissue. During dry salting, an overflow of exudate occurs, which, in the form of brine, is freely drained or may even cover the parts in the container. Due to microbial multiplication possibilities during curing, the product should be kept at around 4 °C, although at lower temperatures the process is longer, it provides greater defence against microbial development. After the curing period, the products are rinsed for excess salt and brine removal and are further air-dried or cold-smoked [3, 12].

The friction penetration for coarse salt is greater than that of table salt as it inflicts superficial lacerations on the meat causing higher osmotic pressures at the point of contact of each coarse salt grain with the meat. Table salt covers the meat more evenly protecting it better from air contact, but on the other hand, it keeps the surface drier, hence penetration occurs more slowly. Its use is preferable when the salting temperature is not extremely low or when trying to salt the middle parts of lean meat. However, salt impurities could act as pro-oxidants increasing fat and myoglobin oxidation in cured meat products [4].

Traditional dry curing does not fit perfectly on large scale production, so curing in tanks using concentrated brines was developed. Immersion curing provides a basic composition for brine, consisting of water and table salt or sweetener, phosphate, erythorbate, and nitrite brine. The brine solution or "pickle" strength is determined by the amount of salt contained in it [7]. For brine immersion or cover pickling, the meat pieces are simply immersed in the brine for a specified period of time. Larger pieces require higher concentration brines [7, 12]. Weight gain is dependent on pH, temperature, meat shape and size, brine/meat ratio, fat content in meat, brine strength, curing duration, and vacuum use [16]. In soaking, the weight gain for pieces of meat range from 3% to 8%. Brine requires permanent control as it may contain large numbers of microorganisms [7].

The multiple-needle injection method became a popular one where the brine is injected mechanically under pressure using needles. In this method, a conveyor belt carries meat under a bank of offset needles that inject brine until the desired target weight is achieved. The main advantages of multiple-needle injection are increased product yield and reduced production time.

Tumbling or massaging of pickle-injected meat cuts speeds up the curing process, facilitates salt-soluble protein extraction, and improves texture, binding, water-holding capacity, and final product yield. The meat massaging enhances NaCl homogenization that in combination with mechanical action, enhances protein solubilization [17, 18].

Combined methods may also be used, where dry-curing is followed by dip-curing or injectable cure with dip or dry cure as well as other combinations, such as

Wiltshire method, a traditional English technique for bacon and ham curing. Historically, the production of Wiltshire cured bacon has involved the injection of whole sides of pork, immersion in a curing brine tank, followed by an extended period of maturation over several days [15, 19].

USE OF NEW METHODS IN SALTED PRODUCTS

The evolution of food processes is driven by changes in consumer preferences and the need to produce safe and high-quality foods. Health awareness and recent studies have confirmed that a high sodium intake (> 2300 mg day^{-1}) [20] could contribute to health problems such as coronary heart disease [21, 22], risk of obesity and central obesity [20], increased atherosclerotic plaque [23], higher risk of urolithiasis and nephrolithiasis [24], chronic kidney disease [25], as well as gastric cancer [26].

Several studies tested meat products reformulations, partially replacing NaCl by KCl, $CaCl_2$, $MgSO_4$ and $MgCl_2$ showing the potential of producing lower sodium content products without compromising their technical and sensory qualities [27 - 29]. However, when used mainly on its own, KCl could damage the sensory acceptance of meat products by fostering bitterness, astringency and a metallic taste [28].

Emerging methods seem to be the best alternative to achieve greater safety and quality in salted meat products [30]. These methods also contribute to production both time and sodium content reduction. In this context, the use of alternative technologies such ultrasound, high hydrostatic pressure, and microwave are mentioned in this chapter.

ULTRASOUND

Ultrasound is a green/sustainable, nonchemical method with various applications in the meat industry, which could be used alone or in combination with other processing and/or preservation methods [31]. Food treatment using ultrasound in a liquid medium, induces acoustic cavitation phenomenon, developing physical forces seen as the main mechanism that changes the exposed materials [30]. Upon reaching their critical size, the bubbles implode, releasing accumulated energy that causes instantaneous and focal temperature increases [32]. The increase in temperature dissipates without causing any substantial increase in the overall temperature of the treated medium. Due to high intensity levels, low-frequency (20-100 kHz) and high intensity (10-1000 W/cm^2) ultrasound waves are able to break intermolecular bonds due to cavitation effect, which modifies the physical properties and chemical reactions [32, 33].

Low-frequency ultrasound could potentially increase meat protein's hydration, accelerate the brining process, significantly increase the salting rate, and allow a homogeneous NaCl distribution [6, 34, 35]. Ultrasound helps reducing brining time without negatively affecting meat characteristics such as quality (colour, texture, expressible moisture, cooking loss), sensory attributes, oxidative stability, and microbial load [36, 37].

Water epithermal cavitation treatment might break the cluster structure's intermolecular hydrogen bonds, thus causing water declustering, and possibly increasing its solvency and dissociation capacity [35]. As for NaCl, the ultrasonic cavitation energy dissociates electrolytes A^+B^-, increasing considerably their solubility. During cavitation treatment, underwater declustering conditions during cavitation treatment, Na^+ and Cl^- ions gain dense solvate shells that prevent their association allowing for their immobilization by water molecules in meat biopolymers – myofibrils of muscle fibres [29, 30].

Processing parameters such as meat section geometry, ultrasound propagation direction, probe size, and distance between probe and sample, influence salt diffusion and distribution in meat [6]. This overflow of factors influencing the effect of ultrasound on meat impairs results comparison [6]. In larger samples processing, it should be considered that mass transport from the solution is proportionate to the external surface and when ultrasound is applied the emitting surface to meat surface ratio as well as intensity distribution around the sample should be taken into account [38]. Inguglia *et al.* observed that the most efficient distance to ensure a higher salt diffusion rate is 0.3 cm [6].

Increasing meat temperature during ultrasound treatment might benefit applications such as cleaning and where thermo sonication is desired, but on the other hand, it might cause protein denaturation favouring bacterial growth [34]. Studies assessing ultrasound effect on lipid oxidation were also evaluated. Analysing the effect of ultrasound on lipid oxidation and meat structure during salting with 6% NaCl an increase on TBARS (Thiobarbituric Acid Reactive Substances) value with longer exposure time and higher ultrasound intensity was observed [39]. Sonicated samples after 7 and 14 days of storage at 4° C presented a more intense fresh meat smell and oily flavour, besides the metallic flavour [40].

Ultrasound could increase the salinity of a food product by reducing the salt crystal size, thereby reducing the total salt added while maintaining the desired salinity [2, 16]. However, such salt tends to dissolve in meat products with moderate to high moisture contents and the crystal size effect disappears [2]. Ultrasound provides an approximately 30% sodium content reduction of cooked hams, as well as decreasing the total release of liquids and augmenting hardness and redness. Moreover, myofibrils micro-fissures and good sensory acceptance were reported [41]. Although the potential use of ultrasound method in speeding up on salting and curing processes is an exciting prospective, the main limitations for the use of powered ultrasound on meat products are related to the limited literature currently available on its effectiveness as a lone meat preservation treatment [16].

HIGH HYDROSTATIC PRESSURE

High-Pressure Food Processing emerged as an innovative non-thermal food preservation method that provides nutritional and sensory advantages over conventional preservation processes. High pressure may affect protein conformation and could lead to its denaturation, aggregation, or gelation [42].

HPP on food processing can be applied in two different forms: a) Isostatic High Pressure (HP) consisting of a HP vessel and an external pressure generating device (Fig. **2**). For HP treatment, the packaged food is placed in a carrier and automatically loaded into the HP vessel and its plugs are closed. The pressure-transmitting media, usually water, is pumped into the vessel from one or both sides.

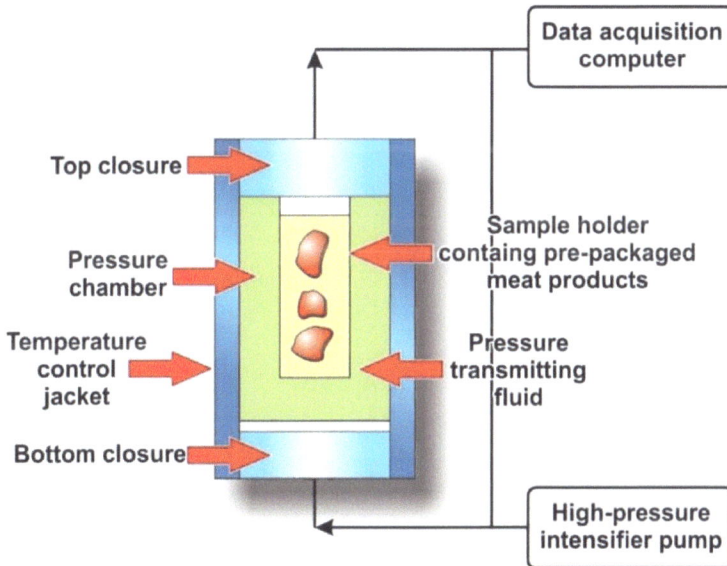

Fig. (2). Schematic diagram of an Isostatic High Pressure (HP) processing system.

Pumping ceases after reaching the desired maximum pressure, and in ideal cases, no further energy input is required to hold the pressure during dwell time [43] and, b) Ultra-High-Pressure Homogenization (UHPH) process; also called Dynamic High-Pressure in the literature (Fig. **3**) processing in ultra-hom-genizers. UHPH is based on the same design principles as the conventional homogenization process used in the dairy industry for reducing the size of fat globules but works at significantly higher pressures (> 200 MPa), resulting in the destruction of microorganisms [44].

An ultra-homogenizer is a single-acting, reciprocating multi-plunger pump with an adjustable homogenizing valve, which produces pressure to micronize and disturb various products. During the UHPH process, the fluid is forced through a narrow gap, where it is submitted to rapid acceleration resulting in shear stress, friction, turbulence, cavitation, and extreme pressure drop [45]. UHPH is often used to prepare food emulsions, producing stable emulsion with fine texture [46]. However, no reports were found in the literature of the use of UHPH in meat products.

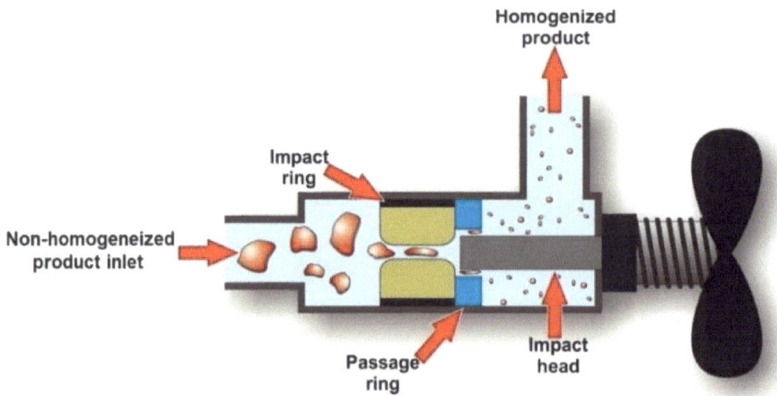

Fig. (3). Schematic diagram of an Ultra-High-Pressure Homogenization (UHPH) processing system.

The high-pressure-processing (HPP) was initially used to control post-cooking recontamination in meat products (whole or sliced), including cooked, dry-cured, marinated, fermented, and raw meat products [9, 47 - 49]. The effect of high hyd-rostatic pressure (HHP) processing combined with nisin application (bio-preservation) was evaluated on the behavior of *Listeria monocytogenes* intentionally inoculated onto the surface of sliced dry-cured ham was evaluated. The results indicated that both hurdles combined provided a wider margin of safety in the control of *L. monocytogenes* during the storage of cured meat products [49].

The feasibility of combined high-pressure carbon dioxide and high-power ultrasound ($HPCO_2$ + HPU) treatment to inactivate *L. monocytogenes* inoculated on the surface of dry cured ham was also investigated. The results revealed that HPU alone was not able to induce any microbial inactivation while $HPCO_2$ + HPU treatment assured a certain level of inactivation. Process conditions of 12 MPa, 35 °C, at 10 W for 5 min assured inactivation to undetectable level of *L. monocytogenes* spiked on the surface of the product with an initial concentration of about 10^9 CFU/g [50].

Bacteria of the genus *Clostridium* are highly worrying for the meat industry and their control is through the use of saltpeter (nitrate) or nitrite, considered carcinogenic. However, no curing salts substitutes have yet been found with promising. HHP has been studied to inhibit this microorganism and satisfactory results have been obtained [51, 52].

Changes in meat macromolecule structures generated by HHP treatment might minimize the impact of sodium reduction on some physicochemical and sensory characteristics [53 - 56]. Clariana *et al.* [57], evaluating cured ham submitted to HPP treatment without any changes in sodium chloride content observed a higher saltiness perception when comparing the treated product with a control. According to the authors, this was due to the weakening of interactions between proteins and sodium, leading to a higher concentration of free ions in the product. However, adverse effects on sensory characteristics have been reported particularly on sliced dry-cured ham. HPP (600 MPa) applied to intact vacuum samples (450 g) minimized the impact of this treatment on appearance, odour, and texture [13].

Various authors evaluated the impact of HHP on physico-chemical, nutritional, and sensory properties of both Iberian and Serrano dry-cured hams [57, 58]. Overall, the results from these studies revealed a decrease in lean colour intensity, pastiness, and juiciness whereas hardness and chewiness increased. HPP enhanced the rancid odour and saltiness of the evaluated dry-cured hams. While in studies by Lorido *et al.* [13] the results revealed that HHP did not have any significant effect on the physico-chemical, fatty acid profile, and sensory properties evaluated by the static techniques (AQD). However, the instrumental colour of both Iberian and Serrano dry-cured hams was affected by the HHP treatment, increasing the parameters L* and b*. Regarding the b* values, this effect was only found to a significant extent ($p < 0.05$) in Serrano samples. Moreover, HPP significantly potentiated saltiness intensity perception. Therefore, HPP treatment could be an alternative to reduce pastiness in salt-reduced dry-cured hams in order to obtain healthier food products with high consumer acceptance [13].

MICROWAVE

Contrary to heating by conduction that characterizes conventional oven cooking or pan frying, electromagnetic microwaves have frequencies between 0.3 and 300 GHz and wavelengths between 1 n and 1 mm that can penetrate food and directly excite specific molecules by oscillating ionic molecules (*e.g.* salt), as well as dipole rotation (water molecules). The overall effects of these movements lead to increased kinetic energy of the molecules, resulting in increased temperature and cooking rate [59].

Fully cooked bacon strips could be cooked using conveyorized, microwave ovens, with a conveyor belt system that holds slices flat to minimize curling during the cooking process [15]. Trials carried out using a pilot-scale industrial microwave-cooking tunnel indicated that the optimum processing condition was found to be 6 kW for 115 s. This condition provided more uniform heating and resulted in lower weight losses than were found in the domestic oven whilst producing bacon of similar organoleptic quality [60]. According to cited authors, it is important to emphasize that cooking is more uniform in an industrial microwave than in a domestic microwave oven, due to uniformity in magnetron position and the movement of material through the microwave system.

FINAL CONSIDERATION

Although salting is one of the oldest techniques used in meat preservation, it is still widely applied in the preparation and preservation of meat products. Salted meat products have high acceptability worldwide. However, in the salted meat products industries, an accurate control of the salting process is complex because of the high degree of heterogeneity in the meat pieces beyond long process time. New processing methods for these products are important to solve these problems. Nowadays, the need for salt reduction of meat products without compromising on quality and safety is a consumer exigence. Ultrasound, HHP and microwave cited in this chapter give good results when applied to salty meat products. However, studies are needed to standardize and enable the use of these new methods processing in the meat industry.

REFERENCES

[1]　M. Dötsch, J. Busch, M. Batenburg, G. Liem, E. Tareilus, R. Mueller, and G. Meijer, "Strategies to reduce sodium consumption: a food industry perspective", *Crit. Rev. Food Sci. Nutr.,* vol. 49, no. 10, pp. 841-851, 2009.
[http://dx.doi.org/10.1080/10408390903044297] [PMID: 19960392]

[2]　E. Desmond, and C. Vasilopoulos, "Reducing salt in meat and poultry products", In: *Reducing Salt in Foods: Practical Strategies.,* D. Kilcast, F. Angus, Eds., 1ˢᵗ ed. Wood Publishing: Cambridge, UK,

2007, pp. 233-251.

[3]   S.M. Lonergan, D.G. Topel, and D.N. Marple, "Fresh and cured meat processing and preservation", In: *The Science of Animal Growth and Meat Technology.*, S.M. Lonergan, D.G. Topel, D.N. Marple, Eds., 2nd ed. Academic Press, 2019, pp. 205-228.
[http://dx.doi.org/10.1016/B978-0-12-815277-5.00013-5]

[4]   M.C. Pardi, I. F. Santos, E.R. Souza, and H.S. Pardi, *Ciência, higiene e tecnologia da carne,* 2nd ed. UFG: Goiânia, 2001.

[5]   S. Damodaran, K.L. Parkin, and O.R. Fennema, *Fennema's Food Chemistry.* 4th ed. Taylor & Francis: UK, 2007.

[6]   E.S. Inguglia, Z. Zhang, C. Burgess, J.P. Kerry, and B.K. Tiwari, "Influence of extrinsic operational parameters on salt diffusion during ultrasound assisted meat curing", *Ultrasonics,* vol. 83, pp. 164-170, 2018.
[http://dx.doi.org/10.1016/j.ultras.2017.03.017] [PMID: 28404320]

[7]   M. Shimokomaki, "Charque, Jerked beef e Carne-de-sol", In: *Atualidades em ciência e tecnologia de carnes,* M. Shimokomaki, R. Olivo, N. N. Terra, B. D. G. M. Franco, Eds., 1st ed. Varela: São Paulo, 2006, pp. 47-62.

[8]   J.M. Lorenzo, "Changes on physico-chemical, textural, lipolysis and volatile compounds during the manufacture of dry-cured foal "cecina"", *Meat Sci.,* vol. 96, no. 1, pp. 256-263, 2014.
[http://dx.doi.org/10.1016/j.meatsci.2013.06.026] [PMID: 23916960]

[9]   B. Rubio, B. Martínez, M.D. García-Cachán, J. Rovira, and I. Jaime, "Effect of high pressure preservation on the quality of dry cured beef 'Cecina de Leon,'", *Innov. Food Sci. Emerg. Technol.,* vol. 8, no. 1, pp. 102-110, 2007.
[http://dx.doi.org/10.1016/j.ifset.2006.08.004]

[10] P.S. Mirade, S. Portanguen, J. Sicard, J. de Souza, A. Musavu Ndob, L.C. Hoffman, T. Goli, and A. Collignan, "Impact of tumbling operating parameters on salt, water and acetic acid transfers during biltong-type meat processing", *J. Food Eng.,* vol. 265, p. 109686, 2020.
[http://dx.doi.org/10.1016/j.jfoodeng.2019.109686]

[11] R. Zhang, M.J. Yoo, J. Gathercole, M.G. Reis, and M.M. Farouk, "Effect of animal age on the nutritional and physicochemical qualities of ovine bresaola", *Food Chem.,* vol. 254, pp. 317-325, 2018.
[http://dx.doi.org/10.1016/j.foodchem.2018.02.031] [PMID: 29548459]

[12] F. Toldrá, and M.C. Aristoy, "Dry - Cured Ham", In: *Handbook of Meat Processing.*, F. Toldrá, Ed., 1st ed. Blackwell Publishing: US, 2010, pp. 351-362.
[http://dx.doi.org/10.1002/9780813820897.ch20]

[13] L. Lorido, M. Estévez, J. Ventanas, and S. Ventanas, "Comparative study between Serrano and Iberian dry-cured hams in relation to the application of high hydrostatic pressure and temporal sensory perceptions", *Lebensm. Wiss. Technol.,* vol. 64, no. 2, pp. 1234-1242, 2015.
[http://dx.doi.org/10.1016/j.lwt.2015.07.029]

[14] L. Purriños, R. Bermúdez, D. Franco, J. Carballo, and J.M. Lorenzo, "Development of volatile compounds during the manufacture of dry-cured "lacón," a Spanish traditional meat product", *J. Food Sci.,* vol. 76, no. 1, pp. C89-C97, 2011.
[http://dx.doi.org/10.1111/j.1750-3841.2010.01955.x] [PMID: 21535660]

[15] C.L. Knipe, and L. Beld, "Bacon Production", In: *Encyclopedia of Meat Sciences,* M. Dikeman, C. Devine, Eds., 2nd ed. Elsevier, 2004, pp. 53-57.

[16] E.S. Inguglia, Z. Zhang, B.K. Tiwari, J.P. Kerry, and C.M. Burgess, "Salt reduction strategies in processed meat products – A review", *Trends Food Sci. Technol.,* vol. 59, pp. 70-78, 2017.
[http://dx.doi.org/10.1016/j.tifs.2016.10.016]

[17] D. Sharedeh, P.S. Mirade, A. Venien, and J.D. Daudin, "Analysis of salt penetration enhancement in

meat tissue by mechanical treatment using a tumbling simulator", *J. Food Eng.,* vol. 166, pp. 377-383, 2015.
[http://dx.doi.org/10.1016/j.jfoodeng.2015.06.023]

[18]   I. Siró, C. Vén, C. Balla, G. Jónás, I. Zeke, and L. Friedrich, "Application of an ultrasonic assisted curing technique for improving the diffusion of sodium chloride in porcine meat", *J. Food Eng.,* vol. 91, no. 2, pp. 353-362, 2009.
[http://dx.doi.org/10.1016/j.jfoodeng.2008.09.015]

[19]   P.R. Sheard, "Wiltshire Sides", *Encycl. Meat Sci.,* vol. 3, pp. 58-63, 2014.
[http://dx.doi.org/10.1016/B978-0-12-384731-7.00109-4]

[20]   X. Zhang, J. Wang, J. Li, Y. Yu, and Y. Song, "A positive association between dietary sodium intake and obesity and central obesity: results from the National Health and Nutrition Examination Survey 1999-2006", *Nutr. Res.,* vol. 55, pp. 33-44, 2018.
[http://dx.doi.org/10.1016/j.nutres.2018.04.008] [PMID: 29914626]

[21]   C. Mizéhoun-Adissoda, D. Houinato, C. Houehanou, T. Chianea, F. Dalmay, A. Bigot, V. Aboyans, P.M. Preux, P. Bovet, and J.C. Desport, "Dietary sodium and potassium intakes: Data from urban and rural areas", *Nutrition,* vol. 33, pp. 35-41, 2017.
[http://dx.doi.org/10.1016/j.nut.2016.08.007] [PMID: 27908548]

[22]   J.M. Barat, and F. Toldrá, "Reducing salt in meat products", In: *Processed Meats: Improving Safety, Nutrition and Quality.,* J.P. Kerry, J.F. Kerry, Eds., 1ˢᵗ ed. Wood Publishing: Cambridge, UK, 2011, p. 724.
[http://dx.doi.org/10.1533/9780857092946.2.331]

[23]   E. Mazza, Y. Ferro, T. Lamprinoudi, C. Gazzaruso, P. Doldo, A. Pujia, and T. Montalcini, "Relationship between high sodium and low PUFA intake and carotid atherosclerosis in elderly women", *Exp. Gerontol,* vol. 108, pp. 256-261, 2018.
[http://dx.doi.org/10.1016/j.exger.2018.05.004]

[24]   B. Afsar, M.C. Kiremit, A.A. Sag, K. Tarim, O. Acar, T. Esen, Y. Solak, A. Covic, and M. Kanbay, "The role of sodium intake in nephrolithiasis: epidemiology, pathogenesis, and future directions", *Eur. J. Intern. Med.,* vol. 35, pp. 16-19, 2016.
[http://dx.doi.org/10.1016/j.ejim.2016.07.001] [PMID: 27444735]

[25]   C.Y. Yoon, J. Noh, J. Lee, Y.K. Kee, C. Seo, M. Lee, M.U. Cha, H. Kim, S. Park, H.R. Yun, S.Y. Jung, J.H. Jhee, S.H. Han, T.H. Yoo, S.W. Kang, and J.T. Park, "High and low sodium intakes are associated with incident chronic kidney disease in patients with normal renal function and hypertension", *Kidney Int.,* vol. 93, no. 4, pp. 921-931, 2018.
[http://dx.doi.org/10.1016/j.kint.2017.09.016] [PMID: 29198468]

[26]   L. D'Elia, G. Rossi, R. Ippolito, F.P. Cappuccio, and P. Strazzullo, "Habitual salt intake and risk of gastric cancer: a meta-analysis of prospective studies", *Clin. Nutr.,* vol. 31, no. 4, pp. 489-498, 2012.
[http://dx.doi.org/10.1016/j.clnu.2012.01.003] [PMID: 22296873]

[27]   J. Zheng, Y. Han, G. Ge, M. Zhao, and W. Sun, "Partial substitution of NaCl with chloride salt mixtures: Impact on oxidative characteristics of meat myofibrillar protein and their rheological properties", *Food Hydrocoll.,* vol. 96, pp. 36-42, 2019.
[http://dx.doi.org/10.1016/j.foodhyd.2019.05.003]

[28]   J.M. Lorenzo, A. Cittadini, R. Bermúdez, P.E. Munekata, and R. Domínguez, "Influence of partial replacement of NaCl with KCl, CaCl2 and MgCl2 on proteolysis, lipolysis and sensory properties during the manufacture of dry-cured lacón", *Food Control,* vol. 55, pp. 90-96, 2015.
[http://dx.doi.org/10.1016/j.foodcont.2015.02.035]

[29]   N. Perisic, N.K. Afseth, R. Ofstad, B. Narum, and A. Kohler, "Characterizing salt substitution in beef meat processing by vibrational spectroscopy and sensory analysis", *Meat Sci.,* vol. 95, no. 3, pp. 576-585, 2013.
[http://dx.doi.org/10.1016/j.meatsci.2013.05.043] [PMID: 23797015]

[30]   A.D. Alarcon-Rojo, L.M. Carrillo-Lopez, R. Reyes-Villagrana, M. Huerta-Jiménez, and I.A. Garcia-Galicia, "Ultrasound and meat quality: A review", *Ultrason. Sonochem.,* vol. 55, pp. 369-382, 2019.
[http://dx.doi.org/10.1016/j.ultsonch.2018.09.016] [PMID: 31027999]

[31]   F. Turantaş, G.B. Kılıç, and B. Kılıç, "Ultrasound in the meat industry: general applications and decontamination efficiency", *Int. J. Food Microbiol.,* vol. 198, pp. 59-69, 2015.
[http://dx.doi.org/10.1016/j.ijfoodmicro.2014.12.026] [PMID: 25613122]

[32]   L.M. Carrillo-lopez, L. Luna-Rodriguez, A.D. Alarcon-Rojo, and M. Huerta-Jimenez, "High intensity ultrasound homogenizes and improves quality of beef longissimus dorsi", *Food Sci. Technol,* vol. 39, no. suppl 1, pp. 332-340, 2019.
[http://dx.doi.org/10.1590/fst.05218]

[33]   A.D. Alarcon-Rojo, H. Janacua, J.C. Rodriguez, L. Paniwnyk, and T.J. Mason, "Power ultrasound in meat processing", *Meat Sci.,* vol. 107, pp. 86-93, 2015.
[http://dx.doi.org/10.1016/j.meatsci.2015.04.015] [PMID: 25974043]

[34]   C.K. McDonnell, P. Allen, G. Duane, C. Morin, E. Casey, and J.G. Lyng, "One-directional modelling to assess the mechanistic actions of power ultrasound on NaCl diffusion in pork", *Ultrason. Sonochem.,* vol. 40, pp. 206-212, 2018.
[http://dx.doi.org/10.1016/j.ultsonch.2017.06.025]

[35]   O. Krasulya, L. Tsirulnichenko, I. Potoroko, V. Bogush, Z. Novikova, A. Sergeev, T. Kuznetsova, and S. Anandan, "The study of changes in raw meat salting using acoustically activated brine", *Ultrason. Sonochem.,* vol. 50, pp. 224-229, 2019.
[http://dx.doi.org/10.1016/j.ultsonch.2018.09.024] [PMID: 30274886]

[36]   C.K. McDonnell, J.G. Lyng, and P. Allen, "The use of power ultrasound for accelerating the curing of pork", *Meat Sci.,* vol. 98, no. 2, pp. 142-149, 2014.
[http://dx.doi.org/10.1016/j.meatsci.2014.04.008] [PMID: 24950083]

[37]   K. S. Ojha, B. K. Tiwari, and C. P. O. Donnell, "Effect of Ultrasound Technology on Food and Nutritional Quality", In: *Advances in Food and Nutrition Research,* F. Toldrá, Ed., vol. 84. Elsevier Inc., 2018, pp. 207-240.
[http://dx.doi.org/10.1016/bs.afnr.2018.01.001]

[38]   J.A. Cárcel, J. Benedito, J. Bon, and A. Mulet, "High intensity ultrasound effects on meat brining", *Meat Sci.,* vol. 76, no. 4, pp. 611-619, 2007.
[http://dx.doi.org/10.1016/j.meatsci.2007.01.022] [PMID: 22061236]

[39]   D.C. Kang, Y.H. Zou, Y.P. Cheng, L.J. Xing, G.H. Zhou, and W.G. Zhang, "Effects of power ultrasound on oxidation and structure of beef proteins during curing processing", *Ultrason. Sonochem.,* vol. 33, pp. 47-53, 2016.
[http://dx.doi.org/10.1016/j.ultsonch.2016.04.024] [PMID: 27245955]

[40]   E. Peña-Gonzalez, A.D. Alarcon-Rojo, I. Garcia-Galicia, L. Carrillo-Lopez, and M. Huerta-Jimenez, "Ultrasound as a potential process to tenderize beef: sensory and technological parameters", *Ultrason. Sonochem,* vol. 53, pp. 134-141, 2018.
[http://dx.doi.org/10.1016/j.ultsonch.2018.12.045] [PMID: 30639205]

[41]   T.L. Barretto, M.A.R. Pollonio, J. Telis-Romero, and A.C. da Silva Barretto, "Improving sensory acceptance and physicochemical properties by ultrasound application to restructured cooked ham with salt (NaCl) reduction", *Meat Sci.,* vol. 145, pp. 55-62, 2018.
[http://dx.doi.org/10.1016/j.meatsci.2018.05.023]

[42]   X.D. Sun, and R.A. Holley, "High hydrostatic pressure effects on the texture of meat and meat products", *J. Food Sci.,* vol. 75, no. 1, pp. R17-R23, 2010.
[http://dx.doi.org/10.1111/j.1750-3841.2009.01449.x] [PMID: 20492191]

[43]   N.K. Rastogi, K.S.M.S. Raghavarao, V.M. Balasubramaniam, K. Niranjan, and D. Knorr, "Opportunities and challenges in high pressure processing of foods", *Crit. Rev. Food Sci. Nutr.,* vol.

47, no. 1, pp. 69-112, 2007.
[http://dx.doi.org/10.1080/10408390600626420] [PMID: 17364696]

[44]   A.X. Roig-Sagués, R.M. Velázquez, P. Montealegre-Agramont, T.J. López-Pedemonte, W.J. Briñez-Zambrano, B. Guamis-López, and M.M. Hernandez-Herrero, "Fat content increases the lethality of ultra-high-pressure homogenization on Listeria monocytogenes in milk", *J. Dairy Sci.,* vol. 92, no. 11, pp. 5396-5402, 2009.
[http://dx.doi.org/10.3168/jds.2009-2495] [PMID. 19841200]

[45]   A.A.L. Tribst, M.A. Franchi, and M. Cristianini, "Ultra-high pressure homogenization treatment combined with lysozyme for controlling Lactobacillus brevis contamination in model system", *Innov. Food Sci. Emerg. Technol.,* vol. 9, no. 3, pp. 265-271, 2008.
[http://dx.doi.org/10.1016/j.ifset.2007.07.012]

[46]   Y. Cha, X. Shi, F. Wu, H. Zou, C. Chang, Y. Guo, M. Yuan, and C. Yu, "Improving the stability of oil-in-water emulsions by using mussel myofibrillar proteins and lecithin as emulsifiers and high-pressure homogenization", *J. Food Eng.,* vol. 258, pp. 1-8, 2019.
[http://dx.doi.org/10.1016/j.jfoodeng.2019.04.009]

[47]   E. Gayán, S.K. Govers, and A. Aertsen, "Impact of high hydrostatic pressure on bacterial proteostasis", *Biophys. Chem.,* vol. 231, pp. 3-9, 2017.
[http://dx.doi.org/10.1016/j.bpc.2017.03.005] [PMID: 28365058]

[48]   R.D. Warner, C.K. McDonnell, A.E.D. Bekhit, J. Claus, R. Vaskoska, A. Sikes, F.R. Dunshea, and M. Ha, "Systematic review of emerging and innovative technologies for meat tenderisation", *Meat Sci.,* vol. 132, pp. 72-89, 2017.
[http://dx.doi.org/10.1016/j.meatsci.2017.04.241] [PMID: 28666558]

[49]   A. Hereu, S. Bover-Cid, M. Garriga, and T. Aymerich, "High hydrostatic pressure and biopreservation of dry-cured ham to meet the Food Safety Objectives for Listeria monocytogenes", *Int. J. Food Microbiol.,* vol. 154, no. 3, pp. 107-112, 2012.
[http://dx.doi.org/10.1016/j.ijfoodmicro.2011.02.027] [PMID: 21411167]

[50]   S. Sara, C. Martina, and F. Giovanna, "High pressure carbon dioxide combined with high power ultrasound processing of dry cured ham spiked with Listeria monocytogenes", *Food Res. Int.,* vol. 66, pp. 264-273, 2014.
[http://dx.doi.org/10.1016/j.foodres.2014.09.024]

[51]   Evelyn and F. V. M. Silva, "High pressure thermal processing for the inactivation of Clostridium perfringens spores in beef slurry", *Innov. Food Sci. Emerg. Technol.,* vol. 33, pp. 26-31, 2016.
[http://dx.doi.org/10.1016/j.ifset.2015.12.021]

[52]   C.J. Doona, F.E. Feeherry, B. Setlow, S. Wang, W. Li, F.C. Nichols, P.K. Talukdar, M.R. Sarker, Y.Q. Li, A. Shen, and P. Setlow, "Effects of high-pressure treatment on spores of Clostridium species", *Appl. Environ. Microbiol.,* vol. 82, no. 17, pp. 5287-5297, 2016.
[http://dx.doi.org/10.1128/AEM.01363-16] [PMID: 27316969]

[53]   A. Grossi, J. Søltoft-Jensen, J.C. Knudsen, M. Christensen, and V. Orlien, "Reduction of salt in pork sausages by the addition of carrot fibre or potato starch and high pressure treatment", *Meat Sci.,* vol. 92, no. 4, pp. 481-489, 2012.
[http://dx.doi.org/10.1016/j.meatsci.2012.05.015] [PMID: 22682686]

[54]   C.M. Crehan, D.J. Troy, and D.J. Buckley, "Effects of salt level and high hydrostatic pressure processing on frankfurters formulated with 1.5 and 2.5% salt", *Meat Sci.,* vol. 55, no. 1, pp. 123-130, 2000.
[http://dx.doi.org/10.1016/S0309-1740(99)00134-5] [PMID: 22060912]

[55]   F. Speroni, N. Szerman, and S.R. Vaudagna, "High hydrostatic pressure processing of beef patties: Effects of pressure level and sodium tripolyphosphate and sodium chloride concentrations on thermal and aggregative properties of proteins", *Innov. Food Sci. Emerg. Technol.,* vol. 23, pp. 10-17, 2014.
[http://dx.doi.org/10.1016/j.ifset.2014.03.011]

[56]   V. Ros-Polski, T. Koutchma, J. Xue, C. Defelice, and S. Balamurugan, "Effects of high hydrostatic pressure processing parameters and NaCl concentration on the physical properties, texture and quality of white chicken meat", *Innov. Food Sci. Emerg. Technol.,* vol. 30, pp. 31-42, 2015.
[http://dx.doi.org/10.1016/j.ifset.2015.04.003]

[57]   M. Clariana, L. Guerrero, C. Sárraga, I. Díaz, Á. Valero, and J.A. García-Regueiro, "Influence of high pressure application on the nutritional, sensory and microbiological characteristics of sliced skin vacuum packed dry-cured ham. Effects along the storage period", *Innov. Food Sci. Emerg. Technol.,* vol. 12, no. 4, pp. 456-465, 2011.
[http://dx.doi.org/10.1016/j.ifset.2010.12.008]

[58]   E. Coll-Brasas, J. Arnau, P. Gou, J.M. Lorenzo, J.V. García-Pérez, and E. Fulladosa, "Effect of high pressure processing temperature on dry-cured hams with different textural characteristics", *Meat Sci.,* vol. 152, pp. 127-133, 2019.
[http://dx.doi.org/10.1016/j.meatsci.2019.02.014] [PMID: 30849689]

[59]   O.P. Soladoye, P. Shand, M.E.R. Dugan, C. Gariépy, J.L. Aalhus, M. Estévez, and M. Juárez, "Influence of cooking methods and storage time on lipid and protein oxidation and heterocyclic aromatic amines production in bacon", *Food Res. Int.,* vol. 99, pp. 660-669, 2017.
[http://dx.doi.org/10.1016/j.foodres.2017.06.029] [PMID: 28784529]

[60]   C. James, K.E. Barlow, S.J. James, and M.J. Swain, "The influence of processing and product factors on the quality of microwave pre-cooked bacon", *J. Food Eng.,* vol. 77, no. 4, pp. 835-843, 2006.
[http://dx.doi.org/10.1016/j.jfoodeng.2005.08.010]

CHAPTER 5

Marinating and Meat Restructuring Technology

Abstract: Meat products such as marinated meats, hamburgers, and nuggets are well accepted by consumers looking for convenience in preparing their meals. These products possess characteristics such as tenderness, juiciness, diversity of flavors, high yield, and also add value to hard-to-sell meats. This chapter will discuss the traditional and innovative meat-producing methods, also called green technologies, currently aimed at making meat products safer, of higher quality, also reducing production costs. Green technologies use either physical, chemical, or biological methods enabling them to perform the same productive tasks more efficiently due to lower electrical power, chemical, and water consumption as well as wastewater production.

Keywords: Batter, Breader, Green technologies, Predust.

INTRODUCTION

Meat is rich in high-quality protein and essential nutrients such as vitamins, minerals, and essential fatty acids [1]. Meat consumption, particularly red meat (beef, pork, and lamb), is dated back to ancient times and remains a dominant lifestyle and nutritionally essential in modern society [2]. Meat quality is influenced by many factors, such as the animal`s breed, genetic aspects, age, gender, pre- and post-slaughter management [3], meat composition just as connective tissue content, streaky or marbled fat, myofibrillar protein components and their myofibrillar contraction level, carcass muscle location, and cooking method [4]. Consumers at the time of purchase choose and evaluate meat cuts based on visual, olfactory, and price perceptions. However, the consumer expects practicality and tenderness during preparation and consumption [5 - 8]. Marinated and restructured processed meat products (breaded or non-breaded) such as seasoned meat cuts, cooked hams, hamburgers and nuggets, meet these consumers' expectations.

Marinated meat is a spicy product produced generally with a whole meat piece with a specific flavor, tenderness, and juiciness [9]. Marinated meat preparation consists of adding brine containing a mixture of additives, spices, and flavorings to the meat. It might occur by immersion, diffusion penetration, injection, or vacuum tumbling [9, 10]. On the other hand, restructured products such as nug-

Daneysa L. Kalschne, Marinês P. Corso & Cristiane Canan

gets are prepared by mechanical disintegration of meat, followed by ingredients mixing and homogenization. The prepared meat batter is placed in containers and inserted in a freezing tunnel [11]. Subsequently, they are nugget-shaped cut, breaded, pre-fried and frozen-stored. For products such as cooked ham - a marinated and restructured meat product - brine is applied to the pork leg either by immersion, injection and/or vacuum tumbling, then the leg cuts are shaped and cooked, resulting in one-piece [12]. It is important to point out that less tender meat remnants from the boning process and difficult-to-market meats could be used for developing marinated and restructured meat products [10]. The subsequent paragraphs describe in more detail the marinated and restructured meat product preparation.

MARINATED MEAT PRODUCTS

Marinating is a process that has been long used that consists of soaking the meat in brine added with simple ingredients, improving the taste and masking some undesirable odors. The brine used in processing is usually a mixture of water, additives, spices, and flavorings and is applied to the meat by dipping, massaging, or tapping, injection for a set period of time. The most demanded functions with marinating are shelf life extension and microbiological, culinary, and technological quality enhancement (jutosity, taste, tenderness, water retention, and mass yield) [13, 14].

Salt is the most important ingredient in marinating as it increases myofibrillar protein solubility and in conjunction with sodium pyrophosphate, sodium hexametaphosphate, and sodium tripolyphosphate enhances the sensory properties of processed meats, increases water retention capacity, assists on the emulsion, color, and flavor stabilization [12], and reduces water loss during the cooking process [15]. The use of acid and alkaline phosphate mixtures provides greater product functionality, contributing to a greater meat water retention capacity, which is of great technological interest, as well as greater juiciness and product slicing easiness. Ingredients such as non-meat proteins, vegetable proteins, hydrocolloids, carrageenan, xanthan, and organic acids mix contribute to the preparation of standardized quality meat products with high water retention capacity, attractive flavor, and low-cost [16].

The marinating methods are primarily responsible for the quality of the final product. The application is essentially static (by meat immersion in brine) or dynamic (using force and friction). Meat immersion in brine is possibly the oldest marinating method, where the migration of ingredients into the myofibrils depends on brine solids concentration and the length of time the meat is dipped [8]. Upon immersion, the salt and acid present in the solution impregnate the tissues, causing a fall in pH, which results in significant hydration of the proteins if the salt content of the solution is less than 0.3 M [13]. For tumbling, the cuts must be carefully chosen due to a high risk of bone breakage or skin separation from certain parts, especially of the chicken, so the tumbler is recommended for small, boneless, skinless pieces of meat, and for other cuts [17]. Injecting equipment that uses pumps to propel the brine through needles into the myofibrillar proteins has been the most used method in the meat industry (Fig. **1**). This equipment provides a spray effect, avoiding brine pockets forming in specific parts of the meat since it maintains brine dispersion in millions of micro-drops with a homogeneous distribution of the marinade in the meat [8]. There is also partial hydration of gel-forming myofibrillar proteins that denature upon heating, leading to meat cuts aggregation [12].

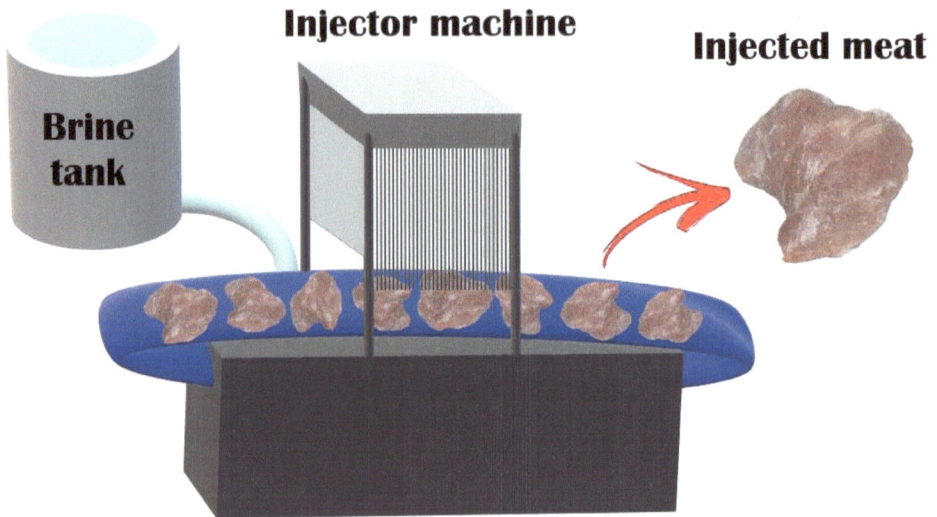

Fig. (1). Injector machine used for meat injection.

The thermal processing (cooking) of ham is carried out in baking ovens until the product reaches an internal temperature of approximately 68 ± 1 °C. At this stage, the killing of most vegetative cells of microorganisms occurs. The survival and/or proliferation of pathogenic microorganisms might occur if the heating treatment is insufficient, which could be dangerous [15].

RESTRUCTURED MEAT PRODUCTS

Consumers nowadays, due to an increasingly busy lifestyle, are seeking products that are quick and easy to prepare, which would make their daily lives easier.

Breaded meat products have been an interesting alternative and their consumption has grown over the years [18]. Noteworthy is restructured and breaded products, such as nuggets, of which preparation is made by muscle disintegration using mechanical processes, followed by mixing of the resulting pieces, later shaped into specific portions [18]. According to Normative Instruction # 6 dated 15 Feb 2001 of the Ministry of Agriculture of Brazil [19], it is established that a breaded meat product is obtained from different butcher's meat, added with ingredients and additives, moulded or not, with an appropriate covering. Breaded meat products are usually pre-fried and complete cooking takes place in a baking oven [18].

Breading meat products also protects the meat from dehydration and cold burning during freezing [20]. One of the purposes for breaded meat product development is using less commercially attractive meat cuts derived from the deboning process and especially those with high-fat content and connective tissue. Fat in breaded meat products undergoes the most quantitative variations. Therefore, the addition of non-meat ingredients in breaded meat product preparation is needed in order to improve the manufacturing conditions and the sensory characteristics of the product. Sodium chloride, phosphates, and polyphosphates are used to increase the ionic strength of the medium and myofibrillar proteins` solubility. Emulsifiers are added to these products to enhance the cohesion and texture of breaded meat products [21].

The preparation of breaded meat products involves meat size reduction (grinding), mixing, moulding, coating (by a specific covering system), pre-frying, and freezing. The grinding process aims at decreasing hardness, since it breaks the raw material into small portions, increasing surface area, and facilitating myofibrillar proteins disposition. The mixing process involves mixing the ingredients (final product formulation) in order to increase the surface area and the breakdown of muscle fiber that favors intracellular components release. Meat and ingredients connection will not take place if protein extraction does not occur, resulting in a product lacking consistency; however, excessive grinding or homogenization will not contribute to good emulsion formation. Subsequently, other ingredients are mixed to achieve uniformity and homogeneity. The moulding is performed by pressing the battered meat into a mould, applying high pressures on the block of the previously frozen meat mixture [18]. After moulding, the meat pieces may or may not be breaded. According to Luvielmo and Dill [22], the traditional meat product breading method is comprised of predust, battering liquid and breading flour (Fig. 2), and the order of these breading steps could be changed.

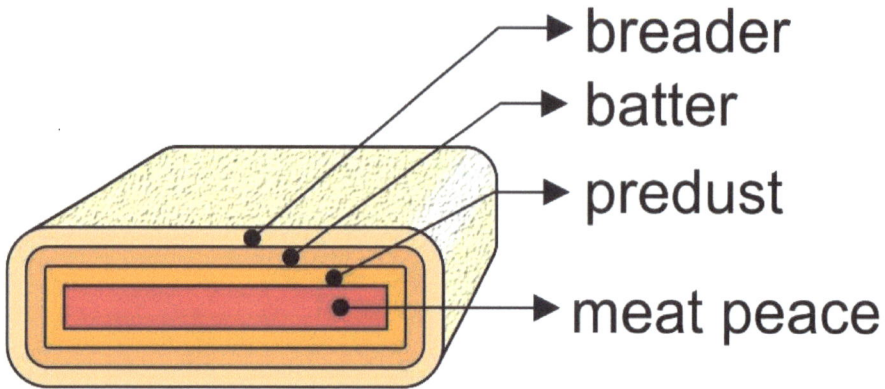

Fig. (2). Breaded meat product coverage systems.

Predust or pre-flouring is the first layer of the coating system, and its main purposes are to foster substrate-batter bonding, absorb moisture from the substrate surface, and maintain the characteristic aroma and flavor [18]. When hydrated, a suspension of solids in liquid forms a complete outer covering layer, acting also as a bonding layer between the substrate and the cover layer.

The batter is essential as it is responsible for the functional and economic characteristics of the product, directly influencing the cover thickness [20]. Subsequently, an external layer of bread is disposed, which promotes a final covering conferring some important characteristics of the product, such as color, apparency, texture, crispy, aroma, and flavor. Different ingredients could be used in batter and breaded layers such as cereal flour, starch, soy protein, chemical leavening, shortening and oil, eggs, milk and whey, carrageenan, xanthan, and a mix of organic acids, flavoring and seasoning, salt, sugar, and hydrocolloids [22, 23].

Breading types vary according to their processing, applications, functions, and end-product-attribute requirements [18] and the grain size is fine (> 60 mesh), medium (20–60 mesh), or large size crumbs (up to 20 mesh size) [24]. The following is a brief description of the main types of breading, according to Chen *et al.* [18]:

a. Flour breading: A cereal based, generally result in a fried coating that provides relatively low surface browning and a very dense coating matrix. Grain-flour breadings can be divided into three subcategories—plain flour breading, classic breading, and original breading—based on inclusion or exclusion of additives or spices. Breading is typically applied to moistened or battered food substrates to influence their color, texture, flavour, and appearance.

b. Cracker meals or cracker crumbs (traditional breadings): Traditional/cracker-type crumbs are usually white or colored breadcrumbs, with minimal or no crust on the surface. This relatively inexpensive crumb with a flat, flake-like structure is easy to use and provides an even surface on the coated product. The browning, achieved during the frying operation, is considered low and the crumbs can be used for full-fry or oven-heated type products. The flakes themselves are fairly dense and give the final product a crunchy texture.

c. American bread crumbs (ABC, home-style bread crumbs): The crumbs come in different sizes and provide a distinct crust and attractive highlighting during the frying operation—a medium to high browning can be achieved. It has a more open structure compared to the flour or the cracker-type crumb, which results in a more crispy texture of the fried product.

d. Japanese bread crumbs (JBC or J-crumb, Oriental-style bread crumbs, or panko): An elongated, typically crust-free crumb with outstanding textural and appearance properties. Since the delicate, three-dimensional structure is fragile, special equipment with minimal friction should be used for its application. The texture of the crumb is fairly open/porous and it is produced as white or colored material, generally used as an outside coating in full fry, oven, and microwave applications.

e. Extruded crumbs: Extruded crumbs are of medium size, strawlike, hard, and open-textured and can be engineered to enhance performance, such as increasing the fryer tolerance of a cracker meal analogue or mimicking the ability to toast like ABC and JBC analogues.

Fig. (**3**) illustrates the appearance of each breading type described below.

The next step in the breaded product processing is pre-frying (Fig. **4**), comprised of soaking the breaded meat product in frying oil for a short period of time at high temperatures (180-200 °C), which vary depending on the product [25], with the internal temperature around 80 °C [26]. Battered and breaded products pre-frying is used to improve mainly crispness, texture, moisture and oil contents, porosity, color, appearance, flavor and nutrition, and partially inhibits product dehydration caused by cold temperatures [11, 22]. Afterwards, the breaded meat product is steamed or simply oven-heated before freezing. Finished frozen product acceptability requires stability during long-term storage without affecting appearance, color, crispness, and flavor.

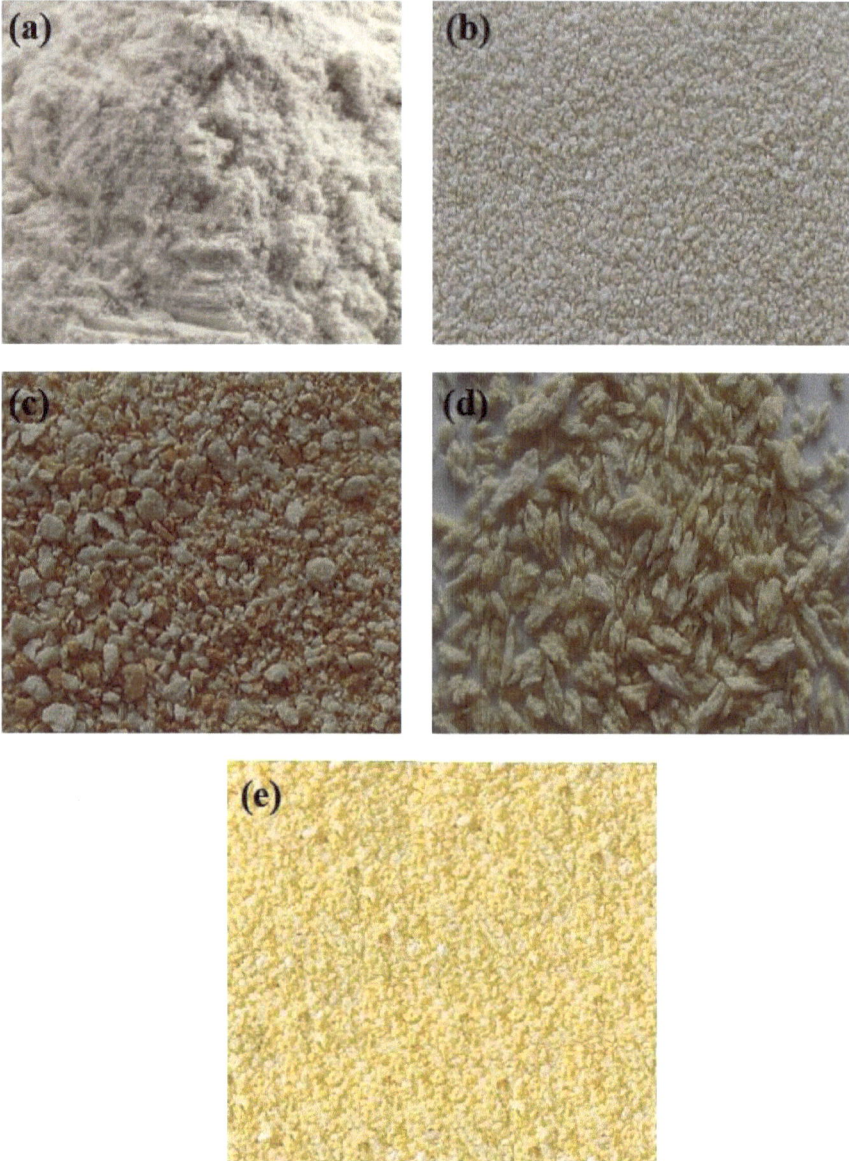

Fig. (3). Breading types: **(a)** flour breading; **(b)** cracker meals or cracker crumb (traditional breadings); **(c)** American bread crumb (ABC, home-style bread crumb); **(d)** Japanese bread crumb (JBC or J-crumb, Oriental-style bread crumb, or panko); and **(e)** extruded crumb.

Frozen coated products are reheated by frying, baking or microwave heating, each yielding different acceptability levels [18, 26]. Baking or microwave is a healthier option but are less acceptable to consumers. In contrast, frying is the most convenient method for consumers since it results in products with more acceptable flavor, golden color, juiciness, and crispness. Oven heating produces a drier product that is moderately acceptable in terms of color, flavor, and crunchiness. Microwave heating, on the other hand, usually results in a pale, soggy product with minimal crunchiness [26, 27]. Since microwaves are a form of energy that manifests itself as heat resulting from direct interaction with the materials and not heat per se, so the microwave oven works basically as a vaporizer [28]. The processing of coated meat products using microwaves should be different from that of conventionally heated products. The coating must adhere well to the food surface but should also allow water vapor to pass through it without being blown off by the internally developed steam pressure [27].

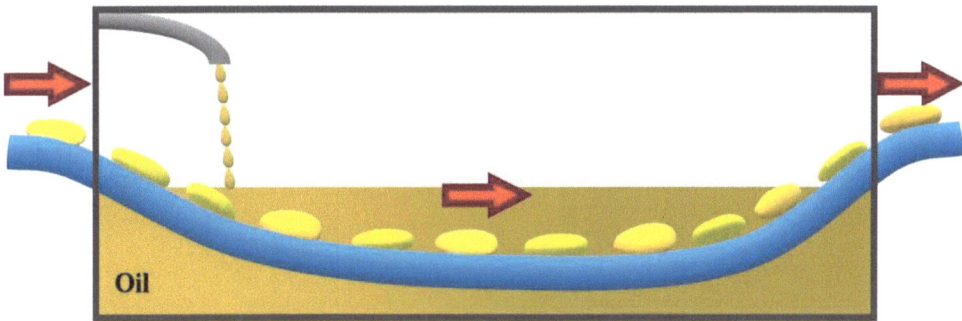

Fig. (4). Pre-frying of meat breaded products.

NEW TECHNOLOGIES

Processed meat manufacturers are constantly looking for new ways to improve quality, safety, shelf-life, and yield of meat products [29]. The meat industry also seeks for product diversity in order to satisfy the consumer [30]. In addition to the traditional methods used, previously mentioned, new technologies are being studied and applied to marinated and/or restructured meat products, coated or uncoated. Such new technologies are high-pressure-processing (HPP), Ultrasound, pulsed light, pulsed ultraviolet light, and microwave, as described below.

HIGH PRESSURE PROCESSING

HPP is an emerging technology of meat product processing due to a number of proven applications, such as reducing salt content, improving uniformity, extending shelf life, and preserving meat products essential, functional, and nutritional characteristics. Despite the initial high investment, pressure treatments consume less energy than thermal processing, a contributing factor to its commercial competitiveness [31, 32].

HPP acts on myofibrillar proteins in a similar manner to salts, therefore, sodium chloride content could be reduced in meat products treated with high-pressure processing not compromising food safety [33]. Furthermore, recent studies assessing the impact of HPP on microbial inactivation in low sodium meat products revealed the importance of salt in meat products for microbial inactivation, offering potential strategies for the development of shelf-stable, low-salt meat products through HPP technology. Table **1** provides a brief overview of some research on marinated meat in which HPP has been applied.

Table 1. High-pressure-processing in marinated meats for microorganism reduction.

Meat products	Pressure (MPa)	Temperature (°C)	Time (min)	Microorganisms	Inactivation ratio ($\log_{10}$ cycles)	Reference
Marinated beef loin	600	31	6	Aerobic, psycrothrophic and LAB	>4	[32]
Cooked ham	400	10	17	*L. monocytogenes*	2.5 at 3.4	[34]
Cooked Ham and Marinated beef loin	600	31	6	*L. monocytogenes, S. enterica, S. aureus, Y. enterocolitica* and *C. Jejuni*	3.5	[35]
Marinated beef (*Longissimus lumborum*)	600	-	5	*L. innocua* and *E. faecium*	6	[36]
Hanwoo beef marinated	550	10	5	Total viable count	0.79 - 1.17	[37]
Marinated pork chops	> 400	3.4 – 6.5	3	Total viable count and *E. coli*	After treatment below the detection limit	[38]

Recent studies evaluated the effect of HPP and the manner meat is prepared and preserved. Marinated herring (*Clupea harengus*) preparation with 4% acetic acid and 500 MPa pressure for 5 or 10 min presented undetectable counting for *Morganella psychrotolerans*, psychrotrophic bacteria, and H_2S-producing bacteria along with three-month storage at 4 °C [39]. However, negative effects on raw meat products' physicochemical characteristics have been reported at higher pressures. Wang *et al.* [40] reported that marinated steaks treatments with pressure ranging from 350 to 450 MPa pressure significantly extended the shelf life without any adverse effect on meat quality. However, studies where higher hydrostatic pressures were used had severe discoloration, impairing trading due to consumer rejection [29, 31].

Fernández *et al.* [31] aiming at minimizing the problem caused by HPP on meat color. Room temperature (20 °C), pressurisation (650 MPa/10 min), and air blast freezing (-30 °C) were compared to air blast freezing plus high pressure at sub-zero temperature (-35 °C). The authors observed that frozen beef pressurised at low temperature showed after-thawing L (luminosity), and b* (blue-yellow parameter) values close to fresh samples. However, these frozen samples presented chromatic parameters similar to unfrozen beef pressurised at room temperature. The authors suggested that freezing protects the meat from pressure color deterioration, with fresh color being recovered after thawing.

Few studies have been conducted in order to evaluate HPP influence on taste. HPP treatment increased the taste compounds content, especially glutamate, sugars, nucleotides, anserine, lactate and creatine, the intensity of umami, sweetness, aftertaste, and overall taste of marinated meat in soy sauce, demonstrating that HPP contributed to improving a palatable meat taste [41]. Physicochemical changes were also observed using HHP. Raw pork chops marinated using HPP at 400 MPa had higher ($p < 0.05$) marinade absorption compared to untreated samples. The increased marinade absorption in pork chops modified the fatty acid composition, increasing ($p < 0.05$) oleic acid levels (C18:1c) [29].

ULTRASOUND

Ultrasound-assisted marinating technology is an alternative to the traditional marinating process, as well as being "green" technology. Acoustic waves applied to solid-liquid systems increase the mass transfer rate by physically breaking down tissues, contributing to greater brine absorption and product uniformity. Ultrasound has been shown to impact tenderness, the most desirable aspect of meat, which is significantly influenced by the water holding capacity of meat [42]. González-González *et al.* [43] concluded that Ultrasound is an alternative to traditional marinating techniques as it delivered a homogeneous brine solutes distribution in the meat. The ultrasonic treatment increases water-binding and water-holding capacity affecting the meat texture when compared to the traditional tumbling method [44, 45].

NaCl diffusion into tissue is accelerated by ultrasonic treatment. However, it should be emphasized that depends on ultrasonic intensity and treatment time. The best results were obtained using 2.5–3.0 Wcm^2 for pork [46]. Extremely high intensities should be avoided as they might result in protein denaturation, with consequent quality loss. On the other hand, extremely low intensities might not be able to induce beneficial structural changes in the tissue, so enhancements in quality attributes cannot be expected. In a study by McDonnell *et al.* [44], a water gain increase in pork meat with 19 W cm^2 for either 10 or 25 min was reported.

The effect of Ultrasound on microorganisms' inhibition in marinated meat has been studied. Ultrasound inhibit *Salmonella sp.* in broiler meat [47], *Escherichia coli* and *Bacillus cereus* in beef [48], and *Listeria monocytogenes* and *Campylobacter jejuni* in pork [49].

PULSED LIGHT AND PULSED ULTRAVIOLET LIGHT

Ready-to-eat foods may pose a safety risk for consumers due to post-processing handing (cutting, slicing, packaging) during which pathogenic bacteria could reach the surface [50]. Pulsed light is used to control pathogenic and spoilage microorganisms, including mould, yeast and viruses in food, surfaces, and packaging improving safety and extending food shelf-life [51, 52]. Photothermal and photochemical mechanisms are responsible for microbial inactivation [33], which makes pulsed light an alternative for microorganism elimination, especially superficial ones found on meat products, such methods use broad-spectrum light pulses and short-time flash [53]. Xenon gas flash lamps emit short-lived, high-power radiation pulses with a spectrum ranging from 200 to 1100 nm, including infrared (IR), visible light (VL) and ultraviolet (UV) [54]. Pulsed ultraviolet light is generated by pulsed UV lamps between 100 and 400 nm, with the potential to decontaminate food products. Studies have shown that pulsed UV-light systems offer cost-effectiveness, high processing speed systems and could be included in processing lines.

High-intensity and UV-pulsed light when applied to meat, fish, and their derivatives caused a significantly microbial reduction in very short treatment time and low environmental impact [53]. However, appropriate treatment conditions evaluation remains necessary (number of flashes, applied voltage, the distance between sample and lamps, light flashes spectral range, surface characteristics, nutrient composition of the sample treated, and the type and amount of microbial contamination) for each food and microorganism separately in order to improve effectiveness and minimize negative attributes that reduce product quality as, in some cases, pulsed light affects negatively on photosensitive compounds and sensory characteristics of food products [52].

Hiero *et al.* [55] studied enhancing the microbial quality and safety of ready-to-eat cooked meat products. Vacuum-packaged ham slices were superficially inoculated with *Listeria monocytogenes*. Pulsed light treatment at 8.4 J/cm^2 reduced *L. monocytogenes* by approximately 2 log CFU/cm^2 in cooked ham slice, not impairing its sensory quality and increasing its shelf-life by 30 days when compared to vacuum-only packaging.

NEAR-INFRARED

Near-infrared (NIR) technology is still poorly studied for meat products. Infrared is invisible electromagnetic energy emitted from objects at extremely high temperatures. The thermal energy could be absorbed by meat products. Ha *et al.* [56] evaluated Near-Infrared pasteurization on ready-to-eat sliced ham and found that L, a*, and b* color parameters were not significantly different ($p > 0.05$) from those of the control. The authors suggested applying NIR heating to control pathogens in ready-to-eat sliced ham without affecting product quality.

FINAL CONSIDERATIONS

New technologies possess a high potential to produce microbiologically safe and high-quality meat products under gentle processes conditions. Process improvement may be achieved by controlled, reproducible studies and interactions between products and used processes, as well as possible undesired changes during the treatment. The implementation of such technologies in the food industry will be dependent on its willingness to innovate, capital invested, operating costs and the cost-benefit involved.

REFERENCES

[1]   C.K. McDonnell, P. Allen, C. Morin, and J.G. Lyng, "The effect of ultrasonic salting on protein and water-protein interactions in meat", *Food Chem.*, vol. 147, pp. 245-251, 2014.
[http://dx.doi.org/10.1016/j.foodchem.2013.09.125] [PMID: 24206713]

[2]   J. Jiang, and Y.L. Xiong, "Natural antioxidants as food and feed additives to promote health benefits and quality of meat products: A review", *Meat Sci.*, vol. 120, pp. 107-117, 2016.
[http://dx.doi.org/10.1016/j.meatsci.2016.04.005] [PMID: 27091079]

[3]   L.E. Jeremiah, "The influence of subcutaneous fat thickness and marbling on beef Palatability and consumer acceptability", *Food Res. Int.*, vol. 29, no. 5–6, pp. 513-520, 1996.
[http://dx.doi.org/10.1016/S0963-9969(96)00049-X]

[4]   R.A. Lawrie, and D.A. Ledward, *Meat Science.* 7th ed. Woodhead Publishing: Cambridge, UK, 2006.

[5]   L.U. Karumendu, Rv. Ven, M.J. Kerr, M. Lanza, and D.L. Hopkins, "Particle size analysis of lamb meat: Effect of homogenization speed, comparison with myofibrillar fragmentation index and its relationship with shear force", *Meat Sci.*, vol. 82, no. 4, pp. 425-431, 2009.
[http://dx.doi.org/10.1016/j.meatsci.2009.02.012] [PMID: 20416692]

[6]   D. Istrati, "The influence of enzymatic tenderization with papain on functional properties of adult beef", *J. Agroaliment. Process. Technol.*, pp. 140-14614, 2008.

[7]   A. Grossi, J. Søltoft-Jensen, J.C. Knudsen, M. Christensen, and V. Orlien, "Reduction of salt in pork sausages by the addition of carrot fibre or potato starch and high pressure treatment", *Meat Sci.*, vol. 92, no. 4, pp. 481-489, 2012.
[http://dx.doi.org/10.1016/j.meatsci.2012.05.015] [PMID: 22682686]

[8]   S.M. Yusop, M.G. O'Sullivan, and J.P. Kerry, "Marinating and enhancement of the nutritional content of processed meat products", In: *Processed Meats: Improving Safety, Nutrition and Quality.* Woodhead Publishing Limited, 2011, pp. 421-449.
[http://dx.doi.org/10.1533/9780857092946.3.421]

[9]   D.P. Smith, and L.L. Young, "Marination pressure and phosphate effects on broiler breast fillet yield, tenderness, and color", *Poult. Sci.,* vol. 86, no. 12, pp. 2666-2670, 2007.
[http://dx.doi.org/10.3382/ps.2007-00144] [PMID: 18029814]

[10]   D.P. Smith, and J.C. Acton, "Marination, cooking and curing of poultry products", In: *Poultry Meat Processing.,* A.R. Sams, Ed., CRC Press LLC: Boca Raton, 2001, pp. 257-279.

[11]   A. Albert, I. Perez-Munuera, A. Quiles, A. Salvador, S.M. Fiszman, and I. Hernando, "Adhesion in fried battered nuggets: Performance of different hydrocolloids as predusts using three cooking procedures", *Food Hydrocoll.,* vol. 23, pp. 1443-1448, 2009.
[http://dx.doi.org/10.1016/j.foodhyd.2008.11.015]

[12]   S. Damodaran, K.L. Parkin, and O.R. Fennema, *Fennema's Food Chemistry.* 4th ed. Taylor & Francis, 2007.

[13]   T. Goli, P. Bohuon, J. Ricci, G. Trystram, and A. Collignan, "Mass transfer dynamics during the acidic marination of turkey meat", *J. Food Eng.,* vol. 104, no. 1, pp. 161-168, 2011.
[http://dx.doi.org/10.1016/j.jfoodeng.2010.12.010]

[14]   T. Goli, J. Ricci, P. Bohuon, S. Marchesseau, and A. Collignan, "Influence of sodium chloride and pH during acidic marination on water retention and mechanical properties of turkey breast meat", *Meat Sci.,* vol. 96, no. 3, pp. 1133-1140, 2014.
[http://dx.doi.org/10.1016/j.meatsci.2013.10.031] [PMID: 24334031]

[15]   M.C. Pardi, I.F. Santos, E.R. Souza, and H.S. Pardi, *Ciência, higiene e tecnologia da carne* 2nd ed. UFG: Goiânia, 2001.

[16]   M. Shimokomaki, "Charque, Jerked beef e Carne-de-sol", In: *Atualidades em ciência e tecnologia de carnes,* M. Shimokomaki, R. Olivo, N. N. Terra, B. D. G. de M. Franco, Eds., 1st ed. Varela: São Paulo, 2006, pp. 47-62.

[17]   "Y. Malila, M. Petracci, S. Benjakul, and W. Visessanguan, "Effect of Tumbling Marination on Marinade Uptake of Chicken Carcass and Parts Quality", *Brazilian J. Poult. Sci.,* vol. 19, no. 1, pp. 61-68, 2017.
[http://dx.doi.org/10.1590/1806-9061-2016-0380]

[18]   R.Y. Chen, Y. Wang, and D. Dyson, "Breadings-What They Are and How They Are Used", In: *Batters and Breadings in Food Processing.,* K. Kulp, R. Loewe, K. Lorenz, J. Gelroth, Eds., 2nd ed. Associate of Cereal Chemists International, 2011, pp. 169-184.
[http://dx.doi.org/10.1016/B978-1-891127-71-7.50015-2]

[19]   Brazil, *Normative Instruction # 6 dated 15 Feb 2001.* Diário Oficial da União: Brasil, 2001.

[20]   F.E. Cunningham, and L.M. Tiede, "Influence of Batter Viscosity on Breading of Chicken Drumsticks", *J. Food Sci.,* vol. 46, pp. 1950-1952, 1981.
[http://dx.doi.org/10.1111/j.1365-2621.1981.tb04527.x]

[21]   H. Dogan, and J.L. Kokini, "Measurement and Interpretation of Batter Rheological Properties", In: *Batters and Breadings in Food Processing.,* K. Kulp, R. Loewe, K. Lorenz, J. Gelroth, Eds., 2nd ed. AACC International, Inc., 2011, pp. 263-299.
[http://dx.doi.org/10.1016/B978-1-891127-71-7.50020-6]

[22]   M. Tamsen, H. Shekarchizadeh, and N. Soltanizadeh, "Evaluation of wheat flour substitution with amaranth flour on chicken nugget properties", *Lebensm. Wiss. Technol.,* vol. 91, pp. 580-587, 2018.
[http://dx.doi.org/https://doi:10.1016/j.lwt.2018.02.001]

[23]   P.K. Mallikarjunan, "Understanding and measuring consumer perceptions of crispness", In: *Texture in Food.,* D. Kilcast, Ed., 1st ed. Woodhead Publishing, 2004, pp. 82-105.
[http://dx.doi.org/10.1533/9781855538362.1.82]

[24]   S. Barbut, "Ingredient Addition and Impacts on Quality, Health, and Consumer Acceptance", In: *Poultry Quality Evaluation.,* M. Petracci, C. Berri, Eds., 1st ed. Woodhead Publishing, 2017, pp. 291-

311.
[http://dx.doi.org/10.1016/B978-0-08-100763-1.00012-X]

[25] T. Gonçalves Albuquerque, M.B.P.P. Oliveira, A. Sanches-Silva, A. Cristina Bento, and H.S. Costa, "The impact of cooking methods on the nutritional quality and safety of chicken breaded nuggets", *Food Funct.,* vol. 7, no. 6, pp. 2736-2746, 2016.
[http://dx.doi.org/10.1039/C6FO00353B] [PMID: 27213579]

[26] V. Jackson, M.W. Schilling, S.M. Falkenberg, T.B. Schmidt, P.C. Coggins, and J.M. Martin, "Quality characteristics and storage stability of baked and fried chicken nuggets formulated with wheat and rice flour", *J. Food Qual.,* vol. 32, no. 6, pp. 760-774, 2009.
[http://dx.doi.org/10.1111/j.1745-4557.2009.00279.x]

[27] R.F. Schiffmann, "Technology of microwavable coated foods", *Batter. Breadings Food Process,* pp. 207-217, 2011.

[28] R. Loewe, "Role of ingredients in batter systems", *Cereal Foods World,* vol. 38, pp. 673-677, 1993.

[29] C.M.O. Neill, M.C. Cruz-romero, G. Duffy, and P. Joe, "Comparative effect of different cooking methods on the physicochemical and sensory characteristics of high pressure processed marinated pork chops", *Innov. Food Sci. Emerg. Technol.,* vol. 54, pp. 19-27, 2019.
[http://dx.doi.org/10.1016/j.ifset.2019.03.005]

[30] H.G.C.L. Gamage, R.K. Mutucumarana, and M.S. Andrew, "Effect of marination method and holding time on physicochemical and sensory characteristics of broiler meat", *J. Agric. Sci.,* vol. 12, no. 3, pp. 172-184, 2017.

[31] P.P. Fernández, P.D. Sanz, A.D. Molina-García, L. Otero, B. Guignon, and S.R. Vaudagna, "Conventional freezing plus high pressure-low temperature treatment: Physical properties, microbial quality and storage stability of beef meat", *Meat Sci.,* vol. 77, no. 4, pp. 616-625, 2007.
[http://dx.doi.org/10.1016/j.meatsci.2007.05.014] [PMID: 22061950]

[32] M. Garriga, N. Grèbol, M.T. Aymerich, J.M. Monfort, and M. Hugas, "Microbial inactivation after high-pressure processing at 600 MPa in commercial meat products over its shelf life", *Innov. Food Sci. Emerg. Technol.,* vol. 5, no. 4, pp. 451-457, 2004.
[http://dx.doi.org/10.1016/j.ifset.2004.07.001]

[33] D.A. Omana, G. Plastow, and M. Betti, "Effect of different ingredients on color and oxidative characteristics of high pressure processed chicken breast meat with special emphasis on use of β-glucan as a partial salt replacer", *Innov. Food Sci. Emerg. Technol.,* vol. 12, no. 3, pp. 244-254, 2011.
[http://dx.doi.org/10.1016/j.ifset.2011.04.007]

[34] B. Marcos, A. Jofré, T. Aymerich, J.M. Monfort, and M. Garriga, "Combined effect of natural antimicrobials and High Pressure Processing to prevent *Listeria monocytogenes* growth after a cold chain break during storage of cooked ham", *Food Control,* vol. 19, no. 1, pp. 76-81, 2008.
[http://dx.doi.org/10.1016/j.foodcont.2007.02.005]

[35] A. Jofré, T. Aymerich, N. Grèbol, and M. Garriga, "Efficiency of high hydrostatic pressure at 600 MPa against food-borne microorganisms by challenge tests on convenience meat products", *Lebensm. Wiss. Technol.,* vol. 42, no. 5, pp. 924-928, 2009.
[http://dx.doi.org/10.1016/j.lwt.2008.12.001]

[36] I. Rodrigues, M.A. Trindade, F.R. Caramit, K. Candoğan, P.R. Pokhrel, and G.V. Barbosa-Cánovas, "Effect of High Pressure Processing on physicochemical and microbiological properties of marinated beef with reduced sodium content", *Innov. Food Sci. Emerg. Technol.,* vol. 38, pp. 328-333, 2016.
[http://dx.doi.org/10.1016/j.ifset.2016.09.020]

[37] Y. An Kim, H. Van Ba, D. Dashdorj, and I. Hwang, "Effect of high-pressure processing on the quality characteristics and shelf-life stability of hanwoo beef marinated with various sauces", *Han-gug Chugsan Sigpum Hag-hoeji,* vol. 38, no. 4, pp. 679-692, 2018.
[PMID: 30206427]

[38]   C.M. O'Neill, M.C. Cruz-Romero, G. Duffy, and J.P. Kerry, "Improving marinade absorption and shelf life of vacuum packed marinated pork chops through the application of High Pressure Processing as a hurdle", *Food Packag. Shelf Life,* vol. 21, p. 100350, 2019.
[http://dx.doi.org/10.1016/j.fpsl.2019.100350]

[39]   I. Ucak, N. Gokoglu, M. Kiessling, S. Toepfl, and C.M. Galanakis, "Inhibitory effects of high pressure treatment on microbial growth and biogenic amine formation in marinated herring (Clupea harengus) inoculated with Morganella psychrotolerans", *Lebensm. Wiss. Technol.,* vol. 99, pp. 50-56, 2018.
[http://dx.doi.org/10.1016/j.lwt.2018.09.058]

[40]   H. Wang, J. Yao, K. Erin, and M. Gänzle, "Effect of pressure on quality and shelf life of marinated beef semitendinosus steaks", *Meat Sci.,* vol. 99, p. 148, 2015.
[http://dx.doi.org/10.1016/j.meatsci.2014.07.011]

[41]   Y. Yang, Y. Ye, Y. Wang, Y. Sun, D. Pan, and J. Cao, "Effect of high pressure treatment on metabolite profile of marinated meat in soy sauce", *Food Chem.,* vol. 240, no. 1, pp. 662-669, 2018.
[http://dx.doi.org/10.1016/j.foodchem.2017.08.006] [PMID: 28946326]

[42]   A.D. Alarcon-Rojo, L.M. Carrillo-Lopez, R. Reyes-Villagrana, M. Huerta-Jiménez, and I.A. Garcia-Galicia, "Ultrasound and meat quality: A review", *Ultrason. Sonochem.,* vol. 55, pp. 369-382, 2018.
[http://dx.doi.org/10.1016/j.ultsonch.2018.09.016] [PMID: 31027999]

[43]   L. González-González, L. Luna-Rodríguez, L. M. Carrillo-López, A. D. Alarcón-Rojo, I. García-Galicia, and R. Reyes-Villagrana, "Ultrasound as an alternative to conventional marination: Acceptability and mass transfer", *J. Food Qual.,* vol. 2017, 2017.
[http://dx.doi.org/10.1155/2017/8675720]

[44]   C.K. McDonnell, J.G. Lyng, and P. Allen, "The use of power Ultrasound for accelerating the curing of pork", *Meat Sci.,* vol. 98, no. 2, pp. 142-149, 2014.
[http://dx.doi.org/10.1016/j.meatsci.2014.04.008] [PMID: 24950083]

[45]   C. Ozuna, A. Puig, J.V. García-Pérez, A. Mulet, and J.A. Cárcel, "Influence of high intensity Ultrasound application on mass transport, microstructure and textural properties of pork meat (Longissimus dorsi) brined at different NaCl concentrations", *J. Food Eng.,* vol. 119, no. 1, pp. 84-93, 2013.
[http://dx.doi.org/10.1016/j.jfoodeng.2013.05.016]

[46]   I. Siró, C. Vén, C. Balla, G. Jónás, I. Zeke, and L. Friedrich, "Application of an ultrasonic assisted curing technique for improving the diffusion of sodium chloride in porcine meat", *J. Food Eng.,* vol. 91, no. 2, pp. 353-362, 2009.
[http://dx.doi.org/10.1016/j.jfoodeng.2008.09.015]

[47]   R. Smith, "Effect of ultrasonic marination on broiler breast meat quality and *salmonella* contamination", *Int. J. Poult. Sci.,* vol. 10, no. 10, pp. 757-759, 2011.
[http://dx.doi.org/10.3923/ijps.2011.757.759]

[48]   D. Kang, Y. Jiang, L. Xing, G. Zhou, and W. Zhang, "Inactivation of *Escherichia coli* O157: H7 and *Bacillus cereus* by power ultrasound during the curing processing in brining liquid and beef", *Food Res. Int,* vol. 102, 2017.
[http://dx.doi.org/10.1016/j.foodres.2017.09.062]

[49]   T. Birk, and S. Knøchel, "Fate of food-associated bacteria in pork as affected by marinade, temperature, and Ultrasound", *J. Food Prot.,* vol. 72, no. 3, pp. 549-555, 2009.
[http://dx.doi.org/10.4315/0362-028X-72.3.549] [PMID: 19343943]

[50]   M. Ganan, E. Hierro, X.F. Hospital, E. Barroso, and M. Fernández, "Use of pulsed light to increase the safety of ready-to-eat cured meat products", *Food Control,* vol. 32, no. 2, pp. 512-517, 2013.
[http://dx.doi.org/10.1016/j.foodcont.2013.01.022]

[51]   V.M. Gómez-López, P. Ragaert, J. Debevere, and F. Devlieghere, "Pulsed light for food decontamination: a review", *Trends Food Sci. Technol.,* vol. 18, no. 9, pp. 464-473, 2007.

[http://dx.doi.org/10.1016/j.tifs.2007.03.010]

[52]   R. Mahendran, K. Ratish Ramanan, F.J. Barba, J.M. Lorenzo, O. López-Fernández, P.E.S. Munekata, S. Roohinejad, A.S. Sant'Ana, and B.K. Tiwari, "Recent advances in the application of pulsed light processing for improving food safety and increasing shelf life", *Trends Food Sci. Technol.,* vol. 88, pp. 67-79, 2019.
[http://dx.doi.org/10.1016/j.tifs.2019.03.010]

[53]   M.L. Bhavya, and H.U. Hebbar, "Pulsed light processing of foods for microbial safety", *Food Qual. Saf.,* vol. 1, no. 3, pp. 187-201, 2017.
[http://dx.doi.org/10.1093/fqsafe/fyx017]

[54]   B. Kramer, J. Wunderlich, and P. Muranyi, "Inactivation of listeria innocua on packaged meat products by pulsed light", *Food Packag. Shelf Life,* vol. 21, 2019.
[http://dx.doi.org/10.1016/j.fpsl.2019.100353]

[55]   E. Hierro, E. Barroso, L. De La Hoz, J.A. Ordóñez, S. Manzano, and M. Fernández, Efficacy of pulsed light for shelf-life extension and inactivation of *Listeria monocytogenes* on ready-to-eat cooked meat products., *Innov. Food Sci. Emerg. Technol.,* vol. 12, no. 3, pp. 275-281, 2011.
[http://dx.doi.org/10.1016/j.ifset.2011.04.006]

[56]   J-W. Ha, S-R. Ryu, and D-H. Kang, "Evaluation of near-infrared pasteurization in controlling *Escherichia coli* O157: H7, *Salmonellaenterica Serovar* typhimurium, and *Listeria monocytogenes* in ready-to-eat sliced ham", *Appl. Environ. Microbiol.,* vol. 78, no. 18, pp. 6458-6465, 2012.
[http://dx.doi.org/10.1128/AEM.00942-12] [PMID: 22773635]

CHAPTER 6

Non-emulsified Sausage Processing Technology

Abstract: Non-emulsified sausages represent a wide group of processed meat products with peculiar characteristics depending on the region of production. In general, they are recognized as thick dough sausage products, stuffing and commercialized in a fresh way, or submitted to smoking or cooking. The fresh non-emulsified sausages are a highly perishable food that cannot be submitted a previous heat treatment before the sale, once the consumer will bake it in the consumption moment. In this context, the high hydrostatic pressure and irradiation are of great interest because only a slight increase in the temperature was observed during the processing step, and can be applied in the finally sealed packages, avoiding re-contamination. Furthermore, for cooked non-emulsified products, the number of possibilities for alternative technologies employment is greater. The high hydrostatic pressure and irradiation can be applied for microbial reduction after cooking step already in the package, and the alternative heating technologies such as ohmic heating, microwave, and radio-frequency can be tested in order to replace the conventional cooking procedures improving the microbiological safety. Thus, the use of emerging technologies represents an interesting point of view in the processing of non-emulsified products, justifying the proper attention of further studies.

Keywords: Electron-beam Irradiation, High hydrostatic pressure, Irradiation, Microwave, Ohmic Heating, Radio-Frequency.

INTRODUCTION

The sausages were recognized as a processed meat product placed inside a wrap. However, there are different groups of sausages, such as the emulsified (thin dough), the dry fermented (thick dough), and the non-emulsified, but proceeded by coarse grinding (thick dough). Fig. (**1**) presents some examples of the three mentioned types of sausages. The non-emulsified sausages may be commercialized as cooked, smoked, or fresh products. The thick dough sausage production includes grinding, seasoning, blending, stuffing, fermentation (applicable only for fermented sausage), smoking (optional), cooking (applicable only for cooked sausage), and packaging. The fresh sausages are highly perishable products if compared with fermented and cooked ones since they are manufactured from fresh ground meat, which is friendly for microbial growth of spoilage and pathogenic microorganisms; moreover, the fresh sausages generally

Daneysa L. Kalschne, Marinês P. Corso & Cristiane Canan

present a high-fat content, favorable for lipid oxidation [1]. In this context, the use of innovative technologies applied for meat and meat products highlights as an alternative, and for sausages, they are primarily used in order to reduce microbial contamination. This chapter gathers information about non-emulsified sausages, less researched in comparison with other sausages types, and some prepositions of alternative technologies that can be applied based on studies already done in other sausages.

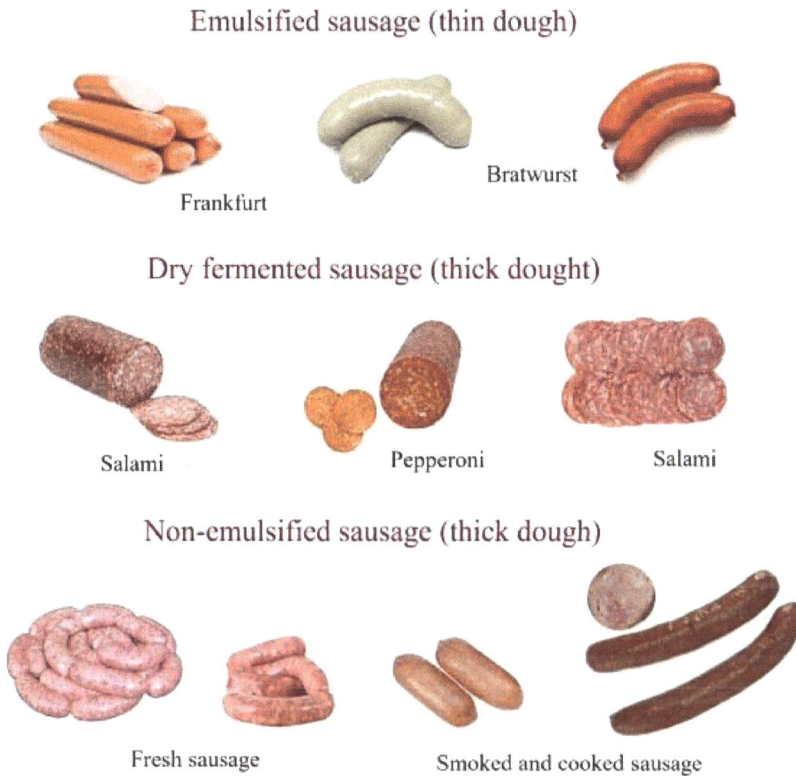

Emulsified sausage (thin dough)

Frankfurt

Bratwurst

Dry fermented sausage (thick dought)

Salami

Pepperoni

Salami

Non-emulsified sausage (thick dough)

Fresh sausage

Smoked and cooked sausage

Fig. (1). Examples of different types of sausages.

USE OF NEW TECHNOLOGIES IN SAUSAGE PRODUCTS

Novel and emerging food processing technologies applied to meat and meat product elaboration have increasingly gained interest. More recently, there has been growing consumer demand for the minimally processed, free from chemical additives and healthier meat, highlighting the development of alternative technologies to improve or replace the conventional heat treatments [2].

The use of steam and hot water in conventional hot treatments for meat or meat products has the disadvantage of slow heat conduction flowing from the heating medium to the thermal center of the product [3]. As the heating is slow, the use of slightly higher temperatures may cause an overheating on the outer surface of the food, effecting the quality of the product, which includes the formation of a free fat layer between the gut and the sausage. In this context, the use of alternative heat treatments represents an interesting solution for the meat industry, especially for uniform sausages that can be heat processed from the inside out, in a standardized way.

HIGH HYDROSTATIC PRESSURE

The high hydrostatic pressure (HHP) is defined by the application of pressure in the range from 100 to 1000 MPa. The emergent technology of HHP has been employed mainly for microorganism reduction in meat and meat products at low or moderate temperature. The highest potential application for meat products might be to pressurize finally sealed packages of contaminated generally sliced products as the color of those products can resist high pressure [4].

According to Balamurugan [5], the effect of HHP (600 MPa for 3 min) was comparable with hot water pasteurization (75 °C internally kept for 15 min) in cooked pork sausage; both pasteurization processes were efficient for preventing the development of both, *Listeria monocytogenes* and lactic acid bacteria, for 35 and 21 days at 4 °C and 10 °C, respectively. Furthermore, both the pasteurization processes were efficient, enabling negative test for *Pseudomonas* spp. and coliforms throughout the 35-day storage at 4 °C and 10 °C.

Similarly, Rubio *et al.* [6] employed the HHP (from 349 to 600 MPa, at 18 °C from 0 to 12.53 min) for *Listeria monocytogenes* inactivation in Spanish chorizo sausage. The authors observed that microorganism reductions increased as the pressure and duration of HHP treatments rose. Moreover, as chorizo presents a low value of water activity, the authors concluded that this parameter seemed to exert a protective effect on the *L. monocytogenes* cells, and pressures below 400 MPa did not lead to significant pathogen reduction.

In the same context, a Spanish blood sausage produced with thick dough and cooked, named Morcilla de Burgos, has been submitted to HHP in order to reduce microbial contamination. The sausage was added to potassium and sodium lactate, and the application of HHP (600 MPa for 10 min) increased the shelf-life of the Morcilla de Burgos by 15 days. Enterobacteria and *Pseudomonas* populations were effectively reduced at all pressures tested (300, 500 and 600 MPa for 10 min). However, for lactic acid bacteria, only the 600 MPa allowed lower counts up to 21 days of storage [7].

The main purpose of using HHP to treat meat products is to improve microbial safety [8]. The effect of HHP on microorganism inactivation is associated with the type of microorganism, pressure level, process temperature and time, pH, and food composition [9]. HPP affects the structure of food constituents, allowing small molecules permeation and filling, adhering it to the surroundings inside larger molecules, such as protein. Furthermore, large molecule chains of protein are lengthened after the HHP decreases to normal pressure. Therefore, the changes may inactivate microorganisms by interrupting the cellular functions responsible for reproduction and survival. HPP can damage microbial membranes, affecting the transport of nutrients and the disposal of cell waste, and cell functions are altered by enzymes inactivation or by the lack of membrane selectivity [10]. In general, Gram-negative bacteria are more sensitive to pressure than Gram-positive ones, but differences in pressure resistance are described among various strains of the same species [9]. In contrast, the destruction of bacterial spores under pressure is more complex, and different combinations of temperature, time, pressure, and cycling treatments can be used to achieve spore inactivation [8].

As HPP can be performed under lower or moderate temperatures, its use can be directed to fresh or fermented non-emulsified sausages, ensuring spoilage or pathogenic microbial reduction without the use of hot treatment, which is impracticable in this group of products. Moreover, the HPP can also be applied for cooked non-emulsified sausages, as long as the cooking step is performed to ensure its characteristics.

IRRADIATION

The electron-beam irradiation can be applied in the low-dose ionization radiation during the treatment of foods in order to eliminate microbial contamination [11]. Regarding meat and meat products, irradiation is classified as an effective technology for inactivating foodborne pathogens, ensuring the safety of meat [12]. The mechanism of microbial inactivation by electron-beam irradiation is associated with high-speed released energy, which targets microbial DNA (nucleic acids), resulting in the microorganism destruction [13, 14].

The microorganism inactivation by electron-beam irradiation can be comparable with gamma rays treatment; however, in practice, these techniques have some differences. The electron-beam irradiation is produced by a machine source based on electricity, while the gamma rays are produced by radioisotopes such as Cobalt-60 [15]. The first advantage, especially if compared with gamma irradiation, is the slight increase in temperature after the treatment and the exemption from radioisotope use [12], which makes consumers afraid about food wholesomeness. Moreover, other difference between both types of irradiation is penetration depth; the penetrating power of gamma rays is much higher than electron-beams. In order to overcome this problem, electron-beam irradiation is applied to both sides of food, improving the penetration depth [14, 15].

The electron-beam irradiation has been applied for emulsified sausages with a great potential for microorganism inactivation. A 2 kGy dose of irradiation during 10 min was effective for *Salmonella* Typhimurium elimination in hot dog sausages and showed no undesirable effects on sensory characteristics [11]. Moreover, the electron-beam irradiation (1, 3, and 8 kGy) applied to a hog and sheep casings presents no influence on physicochemical and sensory properties of emulsion cooked sausage, however, significantly decreased the total aerobic bacteria count; a 3 kGy dose increased the shelf-stability for 5 weeks [16]. In contrast with the considerable decrease of microorganisms, the irradiation has been associated with an increase in the oxidation in meat products [17]. A 0, 3, and 4.5 kGy significantly increased the peroxide value, thiobarbituric acid reactive substances for lipid oxidation, and carbonyl content for protein oxidation in sausages [18]. In order to solve this problem, studies have been conducted combining the electron-beam irradiation with antioxidant sources, generally added in the meat sausage formulation [12, 18].

Practically, no increase in product temperature occurs during electron-beam irradiation application; therefore, it can be used in fresh or fermented non-emulsified sausages. Another advantage is that the irradiation can be applied through in the final package, even though it does not support heat and avoid re-contamination of the product. Additionally, the electron-beam irradiation can also be applied for cooked non-emulsified sausages, ensuring its microbial quality after the cooking step. Fig. (**2**) demonstrates a possible application of electron-beam irradiation in meat products in the final package.

OHMIC HEATING, MICROWAVE, AND RADIO-FREQUENCY COOKING

These three innovative technologies present a good potential for application in meat and meat products; however, for non-emulsified sausages, the use of mentioned technologies is scarce. We believe that the use of these technologies should be tested in non-emulsified cooked sausages in order to replace the traditional cooking step. The use of electrical current to heat foods has been gaining much attraction as a quick and safe heat treatment [4]. This physical principle is the basis of ohmic heating, microwave, and radio-frequency, as detailed below.

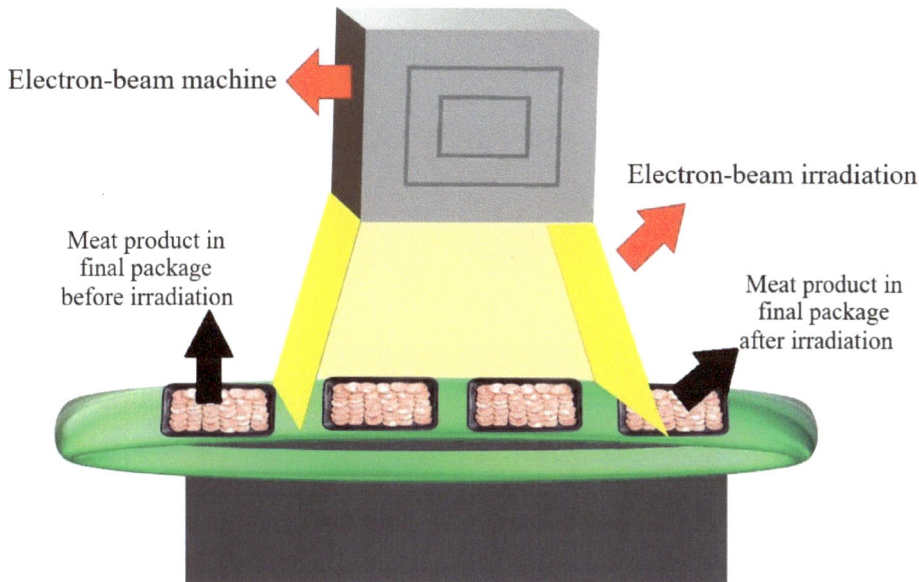

Fig. (2). Examples of different types of sausages.

Ohmic heating is based on the passage of electrical current through a food product having an electrical resistance [4]. For ohmic heating, the electrical conductivity should be the controlling variable; moreover, the efficacy of ohmic heating processing can be influenced by the conductivities of individual components within the food and their behavior and interactions during the heating process [19]. This innovative technology is considered an alternative for ready-to-eat meat products pasteurization inside the final package. Ohmic heating, or electrical resistance heating, is generated internally when an electrical current passes directly through a food product.

Some advantages were associated with ohmic heating, processing time reduction, and uniform temperature distribution. Furthermore, the ohmic heating minimizes alterations in meat texture, color, and appearance while effectively inactivating pathogens [20, 21] (Fig. **3**).

Microwave heating has different applications in the field of food processing, such as cooking, drying and pasteurization [22]. In the microwave heating, the food product forms a dielectric media between the two electrodes and the heating is caused by the internal friction of the polar molecules, especially moisture or water [23].

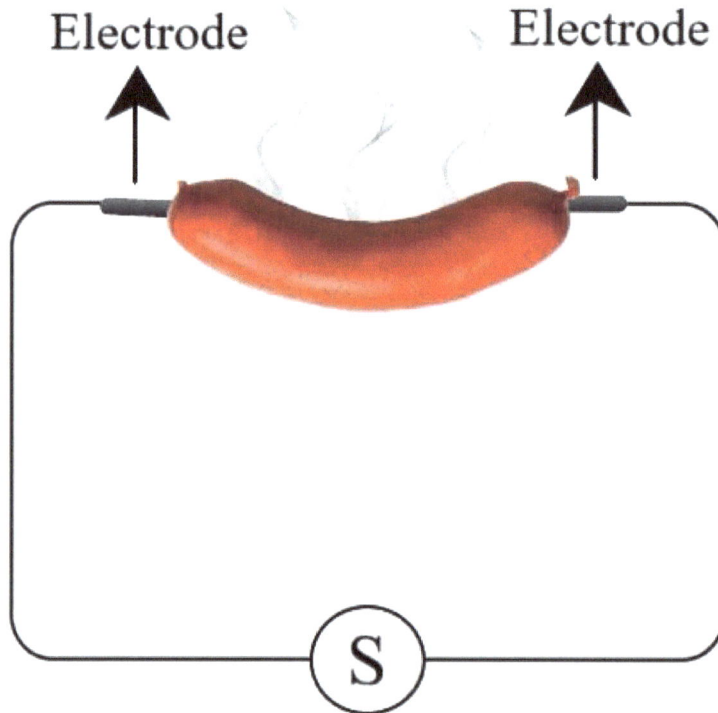

Fig. (3). Scheme of sausage submitted to ohmic heating.

Furthermore, the ionic conduction also contributes to the heating mechanism, associated with the polarization of ions as a result of the back and forth movement of the ionic molecules trying to align themselves with the oscillation field [24]. The frequency of microwave ranges from 300 MHz to 300 GHz; however, for industrial applications, the most commonly used frequencies are 896 MHz and 915 MHz, and 2450 MHz, for commercial, domestic and industrial uses in most countries, respectively [22, 25, 26].

The radio-frequency differs from the microwave as they use the MHz range, while microwave uses the GHz one. The radio-frequency ranges from 10 to 300 MHz, however the band of 13.56, 27.12, and 40.86 MHz have been successfully used for a wide range of applications, including food processing [3, 24, 27 - 29]. Radio-frequency cooking is considered a promising heating technique for meat industry because it can release electromagnetic energy able to provide rapid and uniform heat distribution within the food [3]. During the radio-frequency process, the food product forms a dielectric media between the two electrodes and the heating is promoted by the internal friction of the polar molecules [4]. Due to the radio frequencies, the heating wavelength is 22 to 360 times greater than microwave heating frequencies. It allows greater penetration depth in a product; thus, the wavelength of radio-frequency is more suitable and effective for large-diameter foodstuffs, such as large-caliber sausages [3, 25].

The above-mentioned innovative technologies can also be applied for cooked non-emulsified sausages in order to obtain cook and pasteurization. Although they represent a potential application, its effects in food products should be investigated regarding costs, weight loss, microbial safety, shelf-life, and product sensory characteristics, even including sensory tests with trained and consumers panel.

FINAL CONSIDERATION

The emerging technologies presented in this chapter stand out as significant advances in non-emulsified sausages, especially in order to reduce the microbial contamination. The high hydrostatic pressure and irradiation are of great interest for fresh fermented non-emulsified sausages because only a slight increase in the temperature was observed during the processing step. Moreover, both alternative technologies can be applied in the finally sealed packages, avoiding re-contamination. On the other hand, for cooked non-emulsified sausages, the ohmic heating, microwave, and radio-frequency represent a group of possible technology that can be tested in order to replace the conventional cooking procedures and also improve the microbiological safety. Furthermore, if the conventional cooking procedure is chosen for non-emulsified cooked products, the high hydrostatic pressure and irradiation technologies can be applied in the final package to avoid the multiplication of re-contamination microbiota.

REFERENCES

[1]   C.J. Hugo, and A. Hugo, "Current trends in natural preservatives for fresh sausage products", *Trends Food Sci. Technol.,* vol. 45, no. 1, pp. 12-23, 2015.
        [http://dx.doi.org/10.1016/j.tifs.2015.05.003]

[2]   S. Sazonova, R. Galoburda, and I. Gramatina, "Application of High-pressure Processing for Safety and Shelf-life Quality of Meat – A Review", In: *Baltic Conference on Food Science and Technology* FOODBALT: Latvian, 2017.
        [http://dx.doi.org/10.22616/FoodBalt.2017.001]

[3]   J.H. Chen, Y. Ren, J. Seow, T. Liu, W.S. Bang, and H.G. Yuk, "Intervention technologies for ensuring microbiological safety of meat: Current and future trends", *Compr. Rev. Food Sci. Food Saf.,* vol. 11, no. 2, pp. 119-132, 2012.
        [http://dx.doi.org/10.1111/j.1541-4337.2011.00177.x]

[4]   E. Tornberg, "Engineering processes in meat products and how they influence their biophysical properties", *Meat Sci.,* vol. 95, no. 4, pp. 871-878, 2013.
        [http://dx.doi.org/10.1016/j.meatsci.2013.04.053] [PMID: 23702340]

[5]   S. Balamurugan, P. Inmanee, J. Souza, P. Strange, T. Pirak, and S. Barbut, "Effects of high pressure processing and hot water pasteurization of cooked sausages on inactivation of inoculated Listeria

monocytogenes, natural populations of lactic acid bacteria, pseudomonas spp., and coliforms and their recovery during storage at 4 a", *J. Food Prot.,* vol. 81, no. 8, pp. 1245-1251, 2018.
[http://dx.doi.org/10.4315/0362-028X.JFP-18-024] [PMID: 29969296]

[6]   B. Rubio, A. Possas, F. Rincón, R.M. García-Gímeno, and B. Martínez, "Model for *Listeria monocytogenes* inactivation by high hydrostatic pressure processing in Spanish chorizo sausage", *Food Microbiol.,* vol. 69, pp. 18-24, 2018.
[http://dx.doi.org/10.1016/j.fm.2017.07.012] [PMID: 28941900]

[7]   A.M. Diez, E.M. Santos, I. Jaime, and J. Rovira, "Application of organic acid salts and high-pressure treatments to improve the preservation of blood sausage", *Food Microbiol.,* vol. 25, no. 1, pp. 154-161, 2008.
[http://dx.doi.org/10.1016/j.fm.2007.06.004] [PMID: 17993389]

[8]   H. Simonin, F. Duranton, and M. de Lamballerie, "New insights into the high-pressure processing of meat and meat products", *Compr. Rev. Food Sci. Food Saf.,* vol. 11, no. 3, pp. 285-306, 2012.
[http://dx.doi.org/10.1111/j.1541-4337.2012.00184.x]

[9]   J.C. Cheftel, and E. Dumay, "Effects of high pressure on dairy proteins: a review", *Prog. Biotechnol.,* vol. 13, pp. 299-308, 1996.
[http://dx.doi.org/10.1016/S0921-0423(06)80050-X]

[10]   C.Y. Wang, H.W. Huang, C.P. Hsu, and B.B. Yang, "Recent advances in food processing using high hydrostatic pressure technology", *Crit. Rev. Food Sci. Nutr.,* vol. 56, no. 4, pp. 527-540, 2016.
[http://dx.doi.org/10.1080/10408398.2012.745479] [PMID: 25629307]

[11]   F. Bouzarjomehri, V. Dad, B. Hajimohammadi, S.P. Shirmardi, and A. Yousefi-Ghaleh Salimi, "The effect of electron-beam irradiation on microbiological properties and sensory characteristics of sausages", *Radiat. Phys. Chem.,* vol. 168, p. 2019, 2020.
[http://dx.doi.org/10.1016/j.radphyschem.2019.108524]

[12]   D.G. Lim, K.H. Seol, H.J. Jeon, C. Jo, and M. Lee, "Application of electron-beam irradiation combined with antioxidants for fermented sausage and its quality characteristic", *Radiat. Phys. Chem.,* vol. 77, no. 6, pp. 818-824, 2008.
[http://dx.doi.org/10.1016/j.radphyschem.2008.02.004]

[13]   Z.H. Mohammad, E.A. Murano, R.G. Moreira, and A. Castillo, "Effect of air- and vacuum-packaged atmospheres on the reduction of *Salmonella* on almonds by electron beam irradiation", *Lwt,* vol. 116, pp. 1-5, 2019.
[http://dx.doi.org/10.1016/j.lwt.2019.108389]

[14]   J. Farkas, "Irradiation for better foods", *Trends Food Sci. Technol.,* vol. 17, no. 4, pp. 148-152, 2006.
[http://dx.doi.org/10.1016/j.tifs.2005.12.003]

[15]   S.G. Jeong, and D.H. Kang, "Inactivation of *Escherichia coli* O157:H7, *Salmonella* Typhimurium, and *Listeria monocytogenes* in ready-to-bake cookie dough by gamma and electron beam irradiation", *Food Microbiol.,* vol. 64, pp. 172-178, 2017.
[http://dx.doi.org/10.1016/j.fm.2016.12.017] [PMID: 28213023]

[16]   H.W. Kim, "Effects of electron beam irradiated natural casings on the quality properties and shelf stability of emulsion sausage", *Radiat. Phys. Chem.,* vol. 81, no. 5, pp. 580-583, 2012.
[http://dx.doi.org/10.1016/j.radphyschem.2011.12.050]

[17]   Y.K. Ham, "Effects of irradiation source and dose level on quality characteristics of processed meat products", *Radiat. Phys. Chem.,* vol. 130, pp. 259-264, 2017.
[http://dx.doi.org/10.1016/j.radphyschem.2016.09.010]

[18]   H.M. Badr, and K.A. Mahmoud, "Antioxidant activity of carrot juice in gamma irradiated beef sausage during refrigerated and frozen storage", *Food Chem.,* vol. 127, no. 3, pp. 1119-1130, 2011.
[http://dx.doi.org/10.1016/j.foodchem.2011.01.113] [PMID: 25214104]

[19]   B.M. McKenna, J. Lyng, N. Brunton, and N. Shirsat, "Advances in radio frequency and ohmic heating

of meats", *J. Food Eng.,* vol. 77, no. 2, pp. 215-229, 2006.
[http://dx.doi.org/10.1016/j.jfoodeng.2005.06.052]

[20]   P. Inmanee, P. Kamonpatana, and T. Pirak, "Ohmic heating effects on *Listeria monocytogenes* inactivation, and chemical, physical, and sensory characteristic alterations for vacuum packaged sausage during post pasteurization", *Lwt,* vol. 108, pp. 183-189, 2019.
[http://dx.doi.org/10.1016/j.lwt.2019.03.027]

[21]   N. Shirsat, N.P. Brunton, J.G. Lyng, B. McKenna, and A. Scannell, "Texture, colour and sensory evaluation of a conventionally and ohmically cooked meat emulsion batter", *J. Sci. Food Agric.,* vol. 84, no. 14, pp. 1861-1870, 2004.
[http://dx.doi.org/10.1002/jsfa.1869]

[22]   S. Chandrasekaran, S. Ramanathan, and T. Basak, "Microwave food processing-A review", *Food Res. Int.,* vol. 52, no. 1, pp. 243-261, 2013.
[http://dx.doi.org/10.1016/j.foodres.2013.02.033]

[23]   A.U. Özcan, and H. Bozkurt, "Physical and chemical attributes of a ready-to-eat meat product during the processing: Effects of different cooking methods", *Int. J. Food Prop.,* vol. 18, no. 11, pp. 2422-2432, 2015.
[http://dx.doi.org/10.1080/10942912.2014.982256]

[24]   G-A. Ştefănoiu, E.E. Tănase, A.C. Miteluţ, and M.E. Popa, "Unconventional treatments of food: microwave *vs.* radiofrequency", *Agric. Agric. Sci. Procedia,* vol. 10, pp. 503-510, 2016.
[http://dx.doi.org/10.1016/j.aaspro.2016.09.024]

[25]   J.C. Atuonwu, and S.A. Tassou, "Quality assurance in microwave food processing and the enabling potentials of solid-state power generators: A review", *J. Food Eng.,* vol. 234, pp. 1-15, 2018.
[http://dx.doi.org/10.1016/j.jfoodeng.2018.04.009]

[26]   D. Jain, J. Tang, F. Liu, Z. Tang, and P.D. Pedrow, "Computational evaluation of food carrier designs to improve heating uniformity in microwave assisted thermal pasteurization", *Innov. Food Sci. Emerg. Technol.,* vol. 48, pp. 274-286, 2018.
[http://dx.doi.org/10.1016/j.ifset.2018.06.015]

[27]   R. Villa-Rojas, M.J. Zhu, B.P. Marks, and J. Tang, "Radiofrequency inactivation of Salmonella Enteritidis PT 30 and Enterococcus faecium in wheat flour at different water activities", *Biosyst. Eng.,* vol. 156, pp. 7-16, 2017.
[http://dx.doi.org/10.1016/j.biosystemseng.2017.01.001]

[28]   X. Wei, S.K. Lau, J. Stratton, S. Irmak, and J. Subbiah, "Radiofrequency pasteurization process for inactivation of Salmonella spp. and Enterococcus faecium NRRL B-2354 on ground black pepper", *Food Microbiol.,* vol. 82, pp. 388-397, 2019.
[http://dx.doi.org/10.1016/j.fm.2019.03.007] [PMID: 31027798]

[29]   E.J. Rifna, S.K. Singh, S. Chakraborty, and M. Dwivedi, "Effect of thermal and non-thermal techniques for microbial safety in food powder: Recent advances", *Food Res. Int.,* vol. 126, pp. 1-21, 2019.
[http://dx.doi.org/10.1016/j.foodres.2019.108654] [PMID: 31732065]

CHAPTER 7

Effect of Ultrasound on Fresh Meat Tenderness

Abstract: Ultrasound is considered a "green technology" based on acoustic energy. It is a mechanical, nonionizing, and nonpolluting technology that contributes to the improvement of numerous food industry processes. Several articles published in recent years have reported the effects of ultrasound on beef tenderness. Therefore, we have chosen to write this chapter as we believe that the use of ultrasound could assist the industry to establish effective and efficient processing conditions to improve meat quality and meet consumer demands. In this context, the effect of ultrasound on beef tenderness should be addressed, especially as beef is less tender than poultry, pork and lamb meat.

Keywords: Cavitation, Emerging Technology, Proteins, Proteolytic Enzymes, Ultrasound Waves.

INTRODUCTION

Red meat is considered a prime food for humans due to the quality of its protein and essential nutrients such as vitamins, minerals, and especially essential fatty acids [1]. According to consumers, tenderness is one of the key attributes they look for when buying fresh meat [2 - 5]. However, expectations are often not met as tenderness is influenced by many factors, such as breed, genetic aspects, age, gender and pre and post-slaughter management [6].

In China, Japan, and the USA, beef acceptability is increased with increased marbling [7], which is related to tender meat [8]. In some countries like Brazil, the genetic basis used for beef production comes from animals with a great Zebu genotype contribution, which negatively affects tenderness since these animals have a low marbling score. Thus, alternatives that could contribute to the improvement of meat tenderness are quite relevant. High-power ultrasound is a green emerging technology offering great potential for application in a large number of food industry processes [4, 9 - 12].

In recent years ultrasound has been gaining space in food technology, and various studies have focused on its application to both fresh meat and processed products. The use of ultrasound aims at increasing water retention capacity, accelerating

Daneysa L. Kalschne, Marinês P. Corso & Cristiane Canan

maturation, improving the palatability of non-prime food cuts [13, 14], lowering energy usage during cooking [15], emulsion elaboration [16], increasing marinated products brine absorption (improving tenderness and yield of different meats) [17 - 19], and contributing to small ice crystal formation during freezing (reducing exudate loss during thawing) [20].

Satisfactory results have been obtained from ultrasound application on meat tenderness [12, 21, 22]. Variables such as cut type, size, weight, and physicochemical and functional properties should be considered when developing ultrasound systems to obtain quality products under optimal conditions [10]. This chapter presents the results of the use of ultrasound in fresh beef tenderness contributing to its improvement and its industrial application.

ULTRASOUND PRINCIPLES

Ultrasound technology is based on the use of high-intensity elastic waves to alter the treated media by the adequate nonlinear phenomena exploitation associated with high amplitudes such as radiation pressure, wave distortion, streaming, and cavitation in liquids and dislocation in solids [12]. Its use in the food industry is related to physical, chemical, functional or microbiological changes that may occur in food due to its use, along with increased productivity, yield and/or quality improvement, as well as food safety assurance. It is also noteworthy that it is a clean, fast, non-invasive, non-destructive, and accurate technology [23], with less processing time, lower water consumption, lower energy expenditure, less wastewater and toxic substances production, and safe food production [24, 25].

Ultrasonic waves could be classified into (a) high frequency (2-20 MHz), low intensity (<1 W cm^{-2}) and non-destructive waves applied in non-invasive imaging techniques, sensors and composition and dispersion or concentration analysis of particles in fluids [26]; (b) low frequency (20-100 kHz) and high intensity (10-1000 W cm^{-2}) waves, which break the intermolecular binding due to higher power levels, causing a cavitation effect that alters physical properties and fosters chemical reactions [27]. Cavitation takes place when areas of pressure change occur in the environment with gas bubbles formation, which might or might not implode. The bubbles increase the expanding gases diffusion in the medium, causing alternation between compression waves (positive pressure) and rarefaction one (negative pressure) [27, 28]. Cavitation could be classified according to the mode in which it is generated, namely acoustic, hydrodynamic, optical and particle [29]. However, acoustic and hydrodynamic cavitations are the only ones with suitable intensities to cause physical and chemical changes in food [30]. The acoustic parameters (frequency, intensity, treatment duration, and temperature) determine the extent of the desired sonication result, as well as the characteristics of the sample [14, 31].

The improvement of both food processing and final product by ultrasound is due to the reactions using cavitation presenting shorter reaction times, increased yield, less severe processing conditions use, higher reaction products selectivity, as well as a lower temperature and pressure use, which contribute to the maintenance and integrity of components of interest in food [30].

Ultrasound is applied by ultrasonic wave generating equipment such as ultrasound baths (indirect application) and immersion probes (direct application), which may vary in size, diameter and tip geometry, depending on the need. The differences between the ultrasound bath and the ultrasound probe are depicted in Fig. (**1**).

ULTRASOUND ON BEEF TENDERNESS

Meat palatability depends on quality parameters such as aroma, taste, appearance, juiciness and tenderness, which are all difficult to control as they are associated with cattle breed, genetic crossbreeding, rearing system (pasture or confinement), age, and gender [6]. The importance of quality parameters, such as tenderness, depends on the product and its final destination. However, at the time of purchase, meat tenderness is the most sought-after attribute by consumers [2, 3].

Muscle meat must undergo some biochemical changes to become the final product, and after that, it needs to develop the proper organoleptic peculiarities, including tenderness [32]. The maturation or resolution of *rigor mortis* comprises of changes after cadaveric stiffness development. At that stage a slow muscle relaxation occurs, causing tenderness to meat after 3-4 days of refrigerated storage [8]. Maturation is also a conservation method. It involves storing meat cuts for 15 to 21 days, at above meat freezing point temperature (around 0 °C) [8].

However, if maturation exceeds long periods, undesirable changes such as non-characteristic color pigment and flavor formation might occur. During maturation, the myoglobin iron atom oxidizes losing its oxygen-binding ability (the myoglobin molecule is now called metmyoglobin). In this oxidized condition, the meat turns brown. Although this color is not harmful, it indicates that the meat is no longer fresh, which is a problem for beef marketing. In contrast, the most compromising issue, beyond meat discoloration, is the microorganism development that can lead to green pigments formation, such as sulfomoglobin produced by H_2S-producing bacteria ending in actual putrefaction.

The effect on meat tenderness during maturation is not due to the dissociation of actomyosin formed during *rigor mortis*, but to the action of several meat-endogenous enzymes, especially cathepsins, calpains, and calcium-activated neutral proteinases. The proteolytic action at Z line structure strategic points and thin and thick filaments causes myofibril integrity loss. Although proteolysis does not occur in connective tissue proteins, certain collagen molecules cross-links break, possibly due to lysosomal enzymes action [3]. However, the desired tenderness or expected quality of meat cannot be obtained from maturation, so mechanical, enzymatic and/or chemical tendering methods are used to improve not-well-consumer-accepted meat cuts palatability [5, 20].

Ultrasound has been studied by various researchers as an innovative meat tenderizing technology. Ultrasound could directly induce cell membrane rupture by muscle structure physical weakening, resulting in possibly more tender meat [37]. Another form would be proteolysis activation by releasing lysosomal cathepsins and Ca^{2+} ions from intracellular reserves to activate calpains [31, 33].

Different beef cuts and application conditions for ultrasound were evaluated. Among the cuts considered soft, the *Semitendinosus* [13, 31, 34 - 36], *Pectoralis* [37], *Semimembranosus* [14, 38], *Longissimus thoracis et lumborum* [39 - 41], and *Longissimus dorsi* [9] have been studied. The cuts were done in a transversal or longitudinal direction of muscle fibers, for sizes ranging from 3 x 3 x 3 cm [41] to 7 x 7 x 8 cm [14]. Ultrasound was preferably applied 24 h after slaughter. The use of ultrasound was evaluated using either ultrasound bath [13, 14, 34 - 37, 40] or ultrasound probes [33, 38, 41] at intervals ranging from 15 s [38] to 60 min [36] at different temperatures, and ultrasound frequencies and powers, generally using vacuum-packed samples.

Contrary to what was originally thought, excessive ultrasound time might not improve tenderness. Using an ultrasound bath with a 25.9 kHz frequency for up to 4 min in the *Semitendinosus* muscle, tenderness was improved for both transversal and longitudinal fiber cuts, considering the beef was immediately cooked after applying ultrasound. However, beef tenderness was observed in ultrasound application at longer time intervals [34].

Samples stored at 3 °C for 4 days after the longer ultrasound times application also had no improvement in tenderness, even when cooked at a 70 °C internal temperature. *Pectoralis* muscle sections were treated with high-intensity (20 kHz, 22 W cm^{-2}) for 5 or 10 min at 4 °C and after kept at 2 °C and aged for 1, 6, and 10 days were analyzed. After treatment, a reduction in the beef shear force over the storage period was observed [37], which indicates that power and storage times after ultrasound play an important role in tenderness. Furthermore, the unpackaged *Pectoralis* muscle treated with ultrasound had a lower weight loss than the one processed by other methods.

The *Semitendinosus* muscle presents high connective tissue content. Therefore, the myofibrillar rupture (and possible tenderization) that should occur, are probably masked by the high collagen and elastin contents. Thus, the ultrasound application in periods of more than 10 min was evaluated to act on the connective tissue. It was suggested that any treatment that could change collagen stability due to collagen fibers physical disruption or by further enzymatic degradation of muscle could improve meat quality and texture properties [4]. Firstly, higher exudation rates and lower water retention capacity were observed due to muscle cell disruption and sarcomere shrinkage. Consequently, the treated samples presented a lower weight than the non-treated ones. However, a reduction in shear force values could be noted when evaluating periods of time of up to 60 min, which were caused by changes in connective tissues, and possible collagen fibers melting (convert to gelatin), which has a greater effect on toughness [36].

Semimembranosus muscle treated using an ultrasound bath at 24 h after slaughter (45 kHz, 2 W cm^{-2}, 2 min, 4 °C) and stored at 4 °C showed a shear force reduction 24 h post-treatment, justified by higher water retention capacity, a feature typical of advanced *post mortem* meat. In this case, maturation was advanced without impairing other quality parameters, such as pH and color [13, 14]. Such results support the hypothesis that ultrasound, could, depending on the application conditions accelerate meat cuts maturation process by accelerating proteolysis, inducting improvement in tenderness, followed by a cellular protein structure fragmentation.

Studies using ultrasound probe also yielded conflicting results. For *Longissimus thoracis et lumborum* muscle, ultrasound was tested on hot muscle (2 h after slaughter) and cold boned (24 h after slaughter) [39]. The cuts were submitted to a 20 kHz frequency and a 63 W cm^{-2} intensity for 15 s periods (averaging 10 applications per cut) so that the entire surface was sonicated. Subsequently, the samples were analyzed during the 1 to 14 days maturation period and no significant differences were found between the samples submitted to ultrasound and the control sample one (without any treatment) for shear force measurement, sensory characteristics, collagen solubility, and myofibrillar proteolysis analyses.

For *Semimembranosus* muscle using a 2.6 MHz frequency and 10 W cm^{-2} intensity in the pre-rigor and post-rigor phases, it was observed a short delay at onset *rigor mortis*, sarcomere elongation (12% to 15%), Z-line changes, and increased intracellular calcium (approximately 30%) in the pre-rigor phase, however, no effect was observed on the meat tenderness. For cuts treated in the post-rigor phase, no structural changes were observed but a slight maturation index improvement after 6 days, yet at 14 days this effect was no longer observed. Only structural changes were observed in sections treated in the pre-rigor phase, since ultrasound might contribute to greater calcium release and cathepsin activity before *rigor mortis* resolution [38].

Longissimus semitendinosus muscle treated with an ultrasound probe at room temperature, 24 kHz frequency and 12 W cm^{-2} intensity, for 0, 30, 60, 120, and 240 s increased the meat temperature by 15-30 °C. Afterwards, the samples were matured for up to 8.5 days at 5 °C. Ultrasound reduced meat shear force when used for 60 s or longer. The benefit of ultrasound treatment during storage and maturation has decreased. The quality parameters evaluation indicated a shear force reduction, without affecting pH, color, and weight loss by cooking and exudation [31].

In recent studies, *Longissimus lumborum* muscle was treated by ultrasound using 0.1% papain enzyme solution and incubated at 65 °C for 30 min due to the optimum 65-80 °C papain temperature, and shear force reduction proportional to the ultrasound use duration was observed [41]. Similarly, *Longissimus dorsi* samples sonicated after stored for at least 7 days at 4 °C presented a shear force lower than the control samples (sonicated 24 h *post-mortem*) [9].

Table **1** summarizes some studies cited in this chapter. The aim was to cite the sample preparation as larger or thicker samples could prevent ultrasound from reaching the inside of the meat, contributing to less effective treatment and contradictory results [42].

Table 1. Studies reporting the use of ultrasound in fresh beef tenderization.

Sample size	Ultrasound Application	Negative results	Positive results*	Reference
Semitendinosus **10.0 x 5.0 x 2.54 cm, with approximately 200 g cut longitudinally or transversely the muscle fibers**	Ultrasound bath, 10 °C, 25.9 kHz, ultrasound exposure for 2, 4, and 8 min (longitudinal section); and 2, 4, 8, and 16 min (transversal section) after cooking at 75 °C.	Samples treated for 8 and 16 min obtained higher shear force than control.	Samples treated by 2 and 4 min of ultrasound showed lower shear force than control.	[34]

(Table 1) cont.....

Sample size	Ultrasound Application	Negative results	Positive results*	Reference
Semitendinosus **6.4 × 2.5 × 70.2 cm 2.5 × 5.1 × 10.2 cm**	1) Ultrasound bath, 20 kHz, 1.55 W cm^{-2}, 8, 16, and 24 min, stored at 3 °C for 4 days and cooked at 70 °C 2) Ultrasound bath, 20 kHz, 1.55 W cm^{-2}, 8, 16, and 24 min at 70 °C, and subsequently cooked to internal temperature at 70 °C.	There was no significant effect on shear force decrease in any of the treatments.	Exudation and water loss from cooking were not significant.	[35]
Pectoralis 1) **8.0 x 8.0 x 2.5 cm** 2) **8.0 x 8.0 x 1.3 cm**	1) Ultrasonic bath, 20 kHz, 22 W cm^{-2}, 5 or 10 min, stored at 2 °C and cooked at 70 °C 2) Ultrasound bath, 20 kHz, 22 W cm^{-2}, 5 or 10 min at 70 °C, cooked after ultrasound application to internal temperature at 70 °C.	Decrease in vivid red flesh color.	Tendency to reduce shear force during storage at 2 °C over 10 days and exudation and water loss after cooking were not significant.	[37]
Longissimus thoracis et lumborum **cuts thick with 2.5 cm**	Ultrasound probe, 20 kHz, 63 W cm^{-2}, repeats 15 s until it reaches the entire sample surface.	There was no significant effect on shear force decrease between sonicated and control samples at storage times 1, 3, and 14 days.	There were no sensory changes between the sonicated and control samples.	[39]
Semimenbranosus **(sample size not described)**	Ultrasonic probe, 2.6 MHz, 10 W cm^{-2}; 2 x 15 s.	There was no decrease in shear force when ultrasound was applied to pre-rigor muscle.	When ultrasound was applied post-rigor there was a slight decrease in shear force after 6 days of storage.	[38]
Longissimus lumborum et thoracis and *Semitendinosus* **6.0 x 4.0 x 2.0 cm with approximately 50 g**	Ultrasonic probe, 24 kHz, 12 W cm^{-2}, maximum application time 240 s.	Throughout storage the effect of ultrasound on softness decreases compared to control.	All samples sonicated for 60 s or more showed shear force reduction compared to control at time 0 and 1 day.	[31]

(Table 1) cont.....

Sample size	Ultrasound Application	Negative results	Positive results*	Reference
Semimembranosus **7.0 x 7.0 x 8.0 with approximately 400 g**	Ultrasound bath, 45 kHz, 2 W cm^{-2}, 120 s, 4 ± 1 °C.	Not detailed.	Reduction of shear force by ultrasound when applied transversely to muscle fibers; advance of *rigor mortis.*	[13]
Semimenbranosus **7.0 x 7.0 x 8.0 cm with approximately 400 g**	Ultrasound bath, 45 kHz, 2 W cm^{-2}, 120 s, 4 ± 1 °C.	The sonicated samples presented a slightly less stable color than the control, but without quality loss.	At the storage time of 48 and 72 h the shear force was significantly lower than the control.	[14]
Longissimus lumborum **thick cut with 2.5 cm**	Ultrasound bath, 600 kHz at 48 kPa and 65 kPa.	There was no significant difference between the control sample and the sonicated sample.	Not detailed.	[40]
Semitendinosus **2.5 x 5.0 x 5.0 cm, aproximtelly 100 g**	Ultrasound bath, 40 kHz, 1500 W for 10, 20, 30, 40, 50, and 60 min.	Increased exudate.	Ultrasound time ≥10 min resulted in the lowest shear force value.	[36]
Longissimus lumborum **3 × 3 × 3 cm**	Ultrasonic probe, 20 kHz, 100 W and 300 W for 10, 20, and 30 min.	Not detailed.	Lower shear force was obtained when papain was combined with ultrasound at 100 W for 20 min.	[41]
Longissimus dorsi **4 × 9 × 2.5 cm**	Ultrasound bath, 40 kHz, 11 W cm^{-2}, 4 °C, 60 min (applying 30 min to each side). Treated samples after storage for 0, 7, and 14 days at 4 °C and then were treated with ultrasound.	Ultrasonicated meat presented a greater intensity of metallic taste.	Shear force of sonicated meat was lower than that of control samples at all storage times.	[9]

* Resultados comparados a amostra controle (sem tratamento com ultrassom).

Table **1** presents different application conditions used by the authors highlighting the studies by Got *et al.* [38] and Sikes *et al.* [40], who used high ultrasound frequencies (2.6 MHz and 600 kHz) and observed no shear force reduction in relation to their controls even during storage. This probably occurs due to the damage on cathepsin, calpains, and calcium-activated neutral proteinases [38] caused by ultrasonic waves, reducing proteolysis activation by release cathepsin from lysosomes and Ca^{2+} ions from intracellular reserves to activate calpains [31,

38]. Moreover, the high frequency used may have damaged the meat, causing excessive water holding capacity loss. Jayasooriya *et al.* [31] reported the best hardness reduction resulted from using a frequency between 20 kHz and 45 kHz.

CHANGES DUE TO ULTRASOUND APPLICATION

Applying ultrasound to the meat to increase tenderness might cause other desirable or undesirable modifications. As ultrasound is an emerging technology, many results are contradictory, as mentioned earlier. However, some studies show that ultrasound might depend on the application conditions, alteration in the color, pH, water holding capacity, lipid oxidation, off-flavors occurrence, minority compounds degradation and structural changes in proteins [43]. Some of these alterations are detailed below.

Color Alterations

Color is a key factor in meat quality, as it is the first sensory characteristic evaluated by the consumer. A bright red color in red meat is related to freshness and hence to consumer acceptance [4]. Jayasooriya *et al.* [31] and Stadnik and Dolatowski [14] observed that the L* (luminosity), a* (green-red), and b* (yellow-blue) color parameters were not affected by ultrasound treatment. However, Chang *et al.* [44] noted no ultrasound effect on L* and a*, but parameter b* was significantly lower than the control one when meat was exposed to ultrasound for 30 min. This was due to the high frequency used (45 kHz), especially when compared to the 24 kHz frequency tested by Jayasooriya *et al.* [31]. Pohlman, Dikeman and Zayas [35] used an ultrasound bath with acoustic parameters of 20 kHz and 1.55 W cm^{-2} intensity for 8, 16, or 24 min on *Semitendinosus* and *Biceps femoris* beef muscles. The authors reported no ultrasound effect on L*, a*, and b* parameters. However, by keeping frequency at 20 kHz and increasing the intensity to 22 W cm^{-2} for 5 or 10 min on *Pectoralis* muscle, higher intensity value for L*, lower for a* and higher for b* were observed. Although no difference was observed between the 5 to 10 min range, ultrasound affected the color parameters, regardless of the exposure time used [37].

Caraveo *et al.* [45] applied ultrasound in 1.27 cm thick sections of *Semitendinosus* muscle using an ultrasound bath at 40 kHz frequency| and 11 W cm^{-2} intensity, for 60 and 90 min. The authors observed initially an increase in the L* parameter value and a decrease in the a*parameter value when using ultrasound for 60 or 90 min, but no difference was observed between treated and control samples after 8 storage days. However, meat treated with ultrasound for 90 min had a higher b * parameter value during the whole storage period. Similarly, *Longissimus dorsi* samples sonicated after 7 storage days showed increased L*, reduced a* and saturation value and higher hue angle [9].

Overall, ultrasound produces meat with high luminosity, but with less redness and color saturation and a tendency to an orange hue. Reports have suggested that ultrasound has little effect on meat color since the heat generated is not enough to denature proteins and pigments [9, 40].

pH and Water Holding Capacity

The pH is one of the most important raw meat quality indicators as it directly affects protein stability and functional properties. A pH decrease shrinks the polypeptide chains networks, which decreases the water holding capacity [4]. Cavitation might foster a local increase in temperature and pressure in the region surrounding ultrasound use. These effects could create bubbles that lead to proteins denaturation and the free radicals production that react with the protein side chains, decreasing the acid groups of the proteins [46]. These protein groups could induce proteolytic enzymes and deaminases release, which increase amino acids and basic amines availability [9].

The rapid pH increase (about 0.2 pH units) with ultrasound treatment poses potentially significant implications for the meat tenderness as the pH rate decreases *post-mortem* and the final pH affects cathepsins and calpains activity (the two main enzymes responsible for meat tenderization) [31]. Recently, Peña-Gonzalez *et al.* [9] when evaluating the ultrasound use in *Longissimus dorsi* samples stored for a minimum of 7 days, observed that ultrasonicated meat pH was 0.05 units higher than that of the untreated control sample. Meat pH increase reduces cathepsin activity and increases calpain activity, which is not a problem since calpain plays the dominant role in *post-mortem* muscle proteolysis [3, 47].

Lipid Oxidation and Off-Flavors Alterations

Free radicals creation, hydrogen peroxide production, among others, could be triggered by the chemical effect of cavitation, resulting from local increases in temperature and pressure, shock waves and microstreaming, leading to significant physical and chemical changes in the quality of ultrasound-treated foods [42]. Other possible issues for ultrasound use are nutritional compounds degradation, as well as off-flavors production caused by lipid and/or protein oxidation. During cavitation, water molecules could be broken down into free radicals, intensifying the chemical reactions that occur in the liquid [4]. The high temperatures and pressures during gas bubbles collapse lead to the formation of hydroxyl radicals (OH •) and hydrogen atoms from the dissociation of water molecules in aqueous solutions [48].

Recent studies have shown that highly reactive free radicals in conjunction with other degradation products could induce chain reactions in foods [43]. Kang *et al.* [49] evaluated the ultrasound effect on beef's lipid oxidation and structure during salting with 6% NaCl. The ultrasound was used at different intensities (2.39, 6.23, 11.32, and 20.96 W cm^{-2}), time periods (30, 60, 90, and 120 min) and 20 kHz frequency. It was found that Thiobarbituric Acid Reactive Substances (TBARS) value went up with increased exposure of ultrasound time and intensity. Sonicated samples after 7 and 14 storage days at 4 °C presented, besides the metallic flavor, a more intense fresh meat smell and oily flavor [9].

Protein Structure Alterations

Alterations caused by high-intensity ultrasound might change the rheological and solubility properties of proteins [50]. Gülseren *et al.* [51] reported structural alterations followed by functional ones in bovine serum albumin (BSA) due to the mechanical, thermal and/or chemical effects caused by ultrasound. Ultrasound treatment could modify protein structure, disulfide bond formation, surface hydrophobicity and secondary structure, enhancing proteins solubility and [46] the particle size, which could result in increased intramolecular mobility [51].

Zhang *et al.* [50] reported increased protein solubility using ultrasound when evaluating the myofibrillar protein gels physical-chemical properties. The authors applied a 20 kHz frequency, a 200 to 1000 W power, for 15 min (pulses), generated 88, 117, 150, 173, and 193 W cm^{-2} ultrasound intensities. Ultrasound could change protein structure by proteins molecular unfolding, exposing the hydrophobic groups and inner areas of the molecules, *i.e.* protein surface further polarizes damage to hydrogen and hydrophobic bonds, resulting in greater water-protein interaction [52]. The changes in α-helix and β-sheet structure are greater the more intense the ultrasound conditions used, such as long periods of time and high intensities. When using 2.39 and 6.23 W cm^{-2} ultrasound intensities, it took 60 min to show any significant α-helix and β-sheet content difference between the sonicated sample and the control, with α-helix decrease and β-sheet increase [49].

Microbiological Alterations

Ultrasound used to reduce the meat shear force might also reduce the microbiological count. Several studies have evaluated this effect mainly on meat products with high microbiological contamination. Microbial inhibition is mostly attributed to the creation of intracellular cavitation that causes thinning of cell membranes, heating, and free radical production [24, 53]. Some studies indicate that in case of high microbiological contamination of meat, ultrasound acts at first only on microbial reduction and the difference between the treated and control samples disappears during storage [35]. According to Caraveo *et al.* [45], the use of ultrasound significantly decreased coliforms, mesophilic and psychrophilic bacteria during the storage period. A 90-min ultrasound treatment was responsible for the greatest reduction in the microorganisms mentioned.

FINAL CONSIDERATIONS

As ultrasound is a purely physical and non-invasive technique, it is an alternative to chemical or thermal processing means with great potential to improve beef tenderness. However, the equipment's variability and study parameters make it difficult to establish the ideal application conditions. Ultrasound effects on meat tenderness depend on frequency, intensity and time, as well as on meat samples types.

REFERENCES

[1]   C.K. McDonnell, P. Allen, C. Morin, and J.G. Lyng, "The effect of ultrasonic salting on protein and water-protein interactions in meat", *Food Chem.,* vol. 147, pp. 245-251, 2014.
        [http://dx.doi.org/10.1016/j.foodchem.2013.09.125] [PMID: 24206713]

[2]   D. Istrati, "The influence of enzymatic tenderization with papain on functional properties of adult beef", *J. Agroaliment. Process. Technol,* vol. January 2008, vol. 14, pp. 140-146, 2008.

[3]   D.L. Hopkins, and G.H. Geesink, "Protein Degradation Postmortem and Tenderization", In: *Applied Muscle Biology and Meat Science,* M. Du, R.J. McCormick, Eds., CRC Press: New York, 2009, pp. 149-173.

[4]   A.D. Alarcon-Rojo, L.M. Carrillo-Lopez, R. Reyes-Villagrana, M. Huerta-Jiménez, and I.A. Garcia-Galicia, "Ultrasound and meat quality: A Review", *Ultrason. Sonochem.,* vol. 55, pp. 369-382, 2019.
        [http://dx.doi.org/10.1016/j.ultsonch.2018.09.016] [PMID: 31027999]

[5]   L. Gonzalez-Gonzalez, A.D. Alarcon-Rojo, L.M. Carrillo-Lopez, I.A. Garcia-Galicia, M. Huerta-Jimenez, and L. Paniwnyk, "Does ultrasound equally improve the quality of beef? An insight into longissimus lumborum, infraspinatus and cleidooccipitalis", *Meat Sci.,* vol. 160, pp. 1-11, 2020.
        [http://dx.doi.org/10.1016/j.meatsci.2019.107963] [PMID: 31693966]

[6]   L.E. Jeremiah, "The influence of subcutaneous fat thickness and marbling on beef Palatability and consumer acceptability", *Food Res. Int.,* vol. 29, no. 5–6, pp. 513-520, 1996.
        [http://dx.doi.org/10.1016/S0963-9969(96)00049-X]

[7]   Y. Mao, D.L. Hopkins, Y. Zhang, and X. Luo, "Consumer attitudes to beef and sheep meat in China", *Am. J. Food Nutr.,* vol. 4, no. 2, pp. 30-39, 2016.
        [http://dx.doi.org/10.1016/j.meatsci.2019.107963] [PMID: 31693966]

[8]   J.A. Ordonez, *Tecnologia de Alimentos: Alimentos de Origem Animal.* vol. 2. 1st ed. Artmed: São Paulo, 2004.

[9]   E. Peña-Gonzalez, A.D. Alarcon-Rojo, I. Garcia-Galicia, L. Carrillo-Lopez, and M. Huerta-Jimenez, "Ultrasound as a potential process to tenderize beef: sensory and technological parameters", *Ultrason. Sonochem.,* 2018.
        [http://dx.doi.org/10.1016/j.ultsonch.2018.12.045] [PMID: 30639205]

[10]   K. Knorr, D. Froehling, A. Jaeger, H. Reineke, K. Schlueter, O. Schoessler, and O. Schoessler, *Emerging Technologies for Food Processing,* Springer: New York, 2011. vol.2

[11]   I. Majid, G.A. Nayik, and V. Nanda, "Ultrasonication and food technology: A review", *Cogent Food Agric.,* vol. 1, no. 1, pp. 1-11, 2015.
        [http://dx.doi.org/10.1080/23311932.2015.1071022]

[12] J.A. Gallego-Juárez, G. Rodriguez, V. Acosta, and E. Riera, "Power ultrasonic transducers with extensive radiators for industrial processing", *Ultrason. Sonochem.,* vol. 17, no. 6, pp. 953-964, 2010.
[http://dx.doi.org/10.1016/j.ultsonch.2009.11.006] [PMID: 20022545]

[13] J. Stadnik, Z.J. Dolatowski, and H.M. Baranowska, "Effect of ultrasound treatment on water holding properties and microstructure of beef (m. semimembranosus) during ageing", *Lebensm. Wiss. Technol.,* vol. 41, no. 10, pp. 2151-2158, 2008.
[http://dx.doi.org/10.1016/j.lwt.2007.12.003]

[14] J. Stadnik, and Z.J. Dolatowski, "Influence of sonication on Warner-Bratzler shear force, colour and myoglobin of beef (m. semimembranosus)", *Eur. Food Res. Technol.,* vol. 233, no. 4, pp. 553-559, 2011.
[http://dx.doi.org/10.1007/s00217-011-1550-5]

[15] W.Zou. Yunhe, Kang Dacheng, Liu. Riu, Qi. Jun, and Zhou. Guanghong, Zhang, "Effects of ultrasonic assisted cooking on the chemical profiles of taste and flavor of spiced beef", *Ultrason. Sonochem.,* vol. 46, pp. 36-45, 2018.
[http://dx.doi.org/10.1016/j.ultsonch.2018.04.005]

[16] A.J. Cichoski, M.S. Silva, Y.S.V. Leães, C.C.B. Brasil, C.R. de Menezes, J.S. Barin, R. Wagner, and P.C.B. Campagnol, "Ultrasound: A promising technology to improve the technological quality of meat emulsions", *Meat Sci.,* vol. 148, pp. 150-155, 2019.
[http://dx.doi.org/10.1016/j.meatsci.2018.10.009] [PMID: 30388479]

[17] O. Krasulya, L. Tsirulnichenko, I. Potoroko, V. Bogush, Z. Novikova, A. Sergeev, T. Kuznetsova, and S. Anandan, "The study of changes in raw meat salting using acoustically activated brine", *Ultrason. Sonochem.,* vol. 50, pp. 224-229, 2019.
[http://dx.doi.org/10.1016/j.ultsonch.2018.09.024] [PMID: 30274886]

[18] Y. Zou, H. Shi, P. Xu, D. Jiang, X. Zhang, W. Xu, and D. Wang, "Combined effect of ultrasound and sodium bicarbonate marination on chicken breast tenderness and its molecular mechanism", *Ultrason. Sonochem.,* vol. 59, pp. 1-8, 2019.
[http://dx.doi.org/10.1016/j.ultsonch.2019.104735] [PMID: 31442769]

[19] A. Sergeev, N. Shilkina, V. Tarasov, S. Mettu, O. Krasulya, V. Bogush, and E. Yushina, "The effect of ultrasound treatment on the interaction of brine with pork meat proteins", *Ultrason. Sonochem.,* vol. 61, pp. 1-8, 2020.
[http://dx.doi.org/10.1016/j.ultsonch.2019.104831] [PMID: 31669847]

[20] A.D. Alarcón-Rojo, H. Janacua, J.C. Rodríguez, L. Paniwnyk, and T.J. Mason, "Power ultrasound in meat processing", *Meat Sci.,* vol. 107, pp. 86-93, 2015.
[http://dx.doi.org/10.1016/j.meatsci.2015.04.015] [PMID: 25974043]

[21] T.S. Awad, H.a. Moharram, O.E. Shaltout, D. Asker, and M.M. Youssef, "Applications of ultrasound in analysis, processing and quality control of food: A review", *Food Res. Int.,* vol. 48, no. 2, pp. 410-427, 2012.
[http://dx.doi.org/10.1016/j.foodres.2012.05.004]

[22] L.P. Fallavena, L.D. Ferreira Marczak, and G.D. Mercali, "Ultrasound application for quality improvement of beef Biceps femoris physicochemical characteristics", *Lwt,* vol. 118, pp. 1-7, 2020.
[http://dx.doi.org/10.1016/j.lwt.2019.108817]

[23] C. Gambuteanu, V. Filimon, and P. Alexe, "Effects of ultrasound on technological properties of meat a review", *Food Sci. Technol.,* vol. 14, no. 2, pp. 176-182, 2013.

[24] F. Chemat, Zill-e-Huma, and M.K. Khan, "Applications of ultrasound in food technology: Processing, preservation and extraction", *Ultrason. Sonochem.,* vol. 18, no. 4, pp. 813-835, 2011.
[http://dx.doi.org/10.1016/j.ultsonch.2010.11.023] [PMID: 21216174]

[25] T.L. Barretto, M.A.R. Pollonio, J. Telis-Romero, and A.C. da Silva Barretto, "Improving sensory acceptance and physicochemical properties by ultrasound application to restructured cooked ham with

salt (NaCl) reduction", *Meat Sci.,* vol. 145, pp. 55-62, 2018.
[http://dx.doi.org/10.1016/j.meatsci.2018.05.023] [PMID: 29894848]

[26] L. de L. Alves, A.J. Cichoski, J.S. Barin, C. Rampelotto, and E.C. Durante, "O ultrassom no amaciamento de carnes", *Cienc. Rural,* vol. 43, no. 8, pp. 1522-1528, 2013.
[http://dx.doi.org/10.1590/S0103-84782013000800029]

[27] P. Piyasena, E. Mohareb, and R.C. McKellar, "Inactivation of microbes using ultrasound: a review", *Int. J. Food Microbiol.,* vol. 87, no. 3, pp. 207-216, 2003.
[http://dx.doi.org/10.1016/S0168-1605(03)00075-8] [PMID: 14527793]

[28] J.A. Cárcel, J. Benedito, J. Bon, and A. Mulet, "High intensity ultrasound effects on meat brining", *Meat Sci.,* vol. 76, no. 4, pp. 611-619, 2007.
[http://dx.doi.org/10.1016/j.meatsci.2007.01.022] [PMID: 22061236]

[29] P.R. Gogate, R.K. Tayal, and A.B. Pandit, "Cavitation: A technology on the horizon", *Curr. Sci.,* vol. 91, no. 1, pp. 35-46, 2006.

[30] D.J. McClements, "Advances in the application of ultrasound in food analysis and processing", *Trends Food Sci. Technol.,* vol. 6, pp. 293-299, 1995.
[http://dx.doi.org/10.1016/S0924-2244(00)89139-6]

[31] S.D. Jayasooriya, P.J. Torley, B.R. D'Arcy, and B.R. Bhandari, "Effect of high power ultrasound and ageing on the physical properties of bovine Semitendinosus and Longissimus muscles", *Meat Sci.,* vol. 75, no. 4, pp. 628-639, 2007.
[http://dx.doi.org/10.1016/j.meatsci.2006.09.010] [PMID: 22064027]

[32] A. Lana, and L. Zolla, "Proteolysis in meat tenderization from the point of view of each single protein: A proteomic perspective", *J. Proteomics,* vol. 147, pp. 85-97, 2016.
[http://dx.doi.org/10.1016/j.jprot.2016.02.011] [PMID: 26899368]

[33] J.G. Lyng, P. Allen, and B.M. McKenna, "The effect on aspects of beef tenderness of pre- and post-rigor exposure to a high intensity ultrasound probe", *J. Sci. Food Agric.,* vol. 78, no. 3, pp. 308-314, 1998.
[http://dx.doi.org/10.1002/(SICI)1097-0010(199811)78:3<308::AID-JSFA123>3.0.CO;2-F]

[34] W.D. Smith, NB. Cannon, JE. McKeith, and FK. O'Brien, "Tenderization of semintendinosus muscle using high intensity ultrasound", *Ultrason. Symp.,* pp. 1371-1374, 1991.
[http://dx.doi.org/10.1109/ULTSYM.1991.234038]

[35] F.W. Pohlman, M.E. Dikeman, and J.F. Zayas, "The effect of low-intensity ultrasound treatment on shear properties, color stability and shelf-life of vacuum-packaged beef semitendinosus and biceps femoris muscles", *Meat Sci.,* vol. 45, no. 3, pp. 329-337, 1997.
[http://dx.doi.org/10.1016/S0309-1740(96)00106-4] [PMID: 22061471]

[36] H.J. Chang, Q. Wang, C.H. Tang, and G.H. Zhou, "Effects of Ultrasound Treatment on Connective Tissue Collagen and Meat Quality of Beef Semitendinosus Muscle", *J. Food Qual.,* vol. 38, no. 4, pp. 256-267, 2015.
[http://dx.doi.org/10.1111/jfq.12141]

[37] F.W. Pohlman, M.E. Dikeman, and D.H. Kropf, "Effects of high intensity ultrasound treatment, storage time and cooking method on shear, sensory, instrumental color and cooking properties of packaged and unpackaged beef pectoralis muscle", *Meat Sci.,* vol. 46, no. 1, pp. 89-100, 1997.
[http://dx.doi.org/10.1016/S0309-1740(96)00105-2] [PMID: 22061848]

[38] F. Got, J. Culioli, P. Berge, X. Vignon, T. Astruc, J.M. Quideau, and M. Lethiecq, "Effects of high-intensity high-frequency ultrasound on ageing rate, ultrastructure and some physico-chemical properties of beef", *Meat Sci.,* vol. 51, no. 1, pp. 35-42, 1999.
[http://dx.doi.org/10.1016/S0309-1740(98)00094-1] [PMID: 22061534]

[39] J.G. Lyng, P. Allen, and B. McKenna, "The effects of pre-and post-rigor high-intensity ultrasound treatment on aspects of lamb tenderness", *Lebensm. Wiss. Technol.,* vol. 31, no. 4, pp. 334-338, 1998.

[http://dx.doi.org/10.1006/fstl.1997.0361]

[40] A.L. Sikes, R. Mawson, J. Stark, and R. Warner, "Quality properties of pre- and post-rigor beef muscle after interventions with high frequency ultrasound", *Ultrason. Sonochem.,* vol. 21, no. 6, pp. 2138-2143, 2014.
[http://dx.doi.org/10.1016/j.ultsonch.2014.03.008] [PMID: 24690296]

[41] S. Barekat, and N. Soltanizadeh, "Improvement of meat tenderness by simultaneous application of high-intensity ultrasonic radiation and papain treatment", *Innov. Food Sci. Emerg. Technol.,* vol. 39, pp. 223-229, 2017.
[http://dx.doi.org/10.1016/j.ifset.2016.12.009]

[42] J. Lee, M. Ashokkumar, K. Yasui, T. Tuziuti, T. Kozuka, A. Towata, and Y. Iida, "Development and optimization of acoustic bubble structures at high frequencies", *Ultrason. Sonochem.,* vol. 18, no. 1, pp. 92-98, 2011.
[http://dx.doi.org/10.1016/j.ultsonch.2010.03.004] [PMID: 20452265]

[43] D. Pingret, A.S. Fabiano-Tixier, and F. Chemat, "Degradation during application of ultrasound in food processing: A review", *Food Control,* vol. 31, no. 2, pp. 593-606, 2013.
[http://dx.doi.org/10.1016/j.foodcont.2012.11.039]

[44] H.J. Chang, X.L. Xu, G.H. Zhou, C.B. Li, and M. Huang, "Effects of characteristics changes of collagen on meat physicochemical properties of beef semitendinosus muscle during ultrasonic processing", *Food Bioprocess Technol.,* vol. 5, no. 1, pp. 285-297, 2012.
[http://dx.doi.org/10.1007/s11947-009-0269-9]

[45] O. Caraveo, A.D. Alarcon-Rojo, A. Renteria, E. Santellano, and L. Paniwnyk, "Physicochemical and microbiological characteristics of beef treated with high-intensity ultrasound and stored at 4 °C", *J. Sci. Food Agric.,* vol. 95, no. 12, pp. 2487-2493, 2015.
[http://dx.doi.org/10.1002/jsfa.6979] [PMID: 25363831]

[46] A. Amiri, P. Sharifian, and N. Soltanizadeh, "Application of ultrasound treatment for improving the physicochemical, functional and rheological properties of myofibrillar proteins", *Int. J. Biol. Macromol.,* vol. 111, pp. 139-147, 2018.
[http://dx.doi.org/10.1016/j.ijbiomac.2017.12.167] [PMID: 29307807]

[47] J.M. Hopkins, "D L; Thompson, "Factors contributing to proteolysis and disruption of myofibrillar proteins and the impact on tenderisation in beef and sheep meat", *Aust. J. Agric. Res.,* vol. 53, pp. 149-166, 2002.
[http://dx.doi.org/10.1071/AR01079]

[48] P. Riesz, and T. Kondo, "Free radical formation induced by ultrasound and its biological implications", *Free Radic. Biol. Med.,* vol. 13, no. 3, pp. 247-270, 1992.
[http://dx.doi.org/10.1016/0891-5849(92)90021-8] [PMID: 1324205]

[49] D.C. Kang, Y.H. Zou, Y.P. Cheng, L.J. Xing, G.H. Zhou, and W.G. Zhang, "Effects of power ultrasound on oxidation and structure of beef proteins during curing processing", *Ultrason. Sonochem.,* vol. 33, pp. 47-53, 2016.
[http://dx.doi.org/10.1016/j.ultsonch.2016.04.024] [PMID: 27245955]

[50] Z. Zhang, J.M. Regenstein, P. Zhou, and Y. Yang, "Effects of high intensity ultrasound modification on physicochemical property and water in myofibrillar protein gel", *Ultrason. Sonochem.,* vol. 34, pp. 960-967, 2017.
[http://dx.doi.org/10.1016/j.ultsonch.2016.08.008] [PMID: 27773327]

[51] I. Gülseren, D. Güzey, B.D. Bruce, and J. Weiss, "Structural and functional changes in ultrasonicated bovine serum albumin solutions", *Ultrason. Sonochem.,* vol. 14, no. 2, pp. 173-183, 2007.
[http://dx.doi.org/10.1016/j.ultsonch.2005.07.006] [PMID: 16959528]

[52] Q.H. Wen, Z.C. Tu, L. Zhang, H. Wang, and H.X. Chang, "Effect of High Intensity Ultrasound on the Gel and Structural Properties of Ctenopharyngodon idellus Myofibrillar Protein", *J. Food Biochem.,* vol. 41, no. 1, pp. 1-10, 2017.

[http://dx.doi.org/10.1111/jfbc.12288]

[53]   F. Turantaş, G.B. Kılıç, and B. Kılıç, "Ultrasound in the meat industry: general applications and decontamination efficiency", *Int. J. Food Microbiol.,* vol. 198, pp. 59-69, 2015.
[http://dx.doi.org/10.1016/j.ijfoodmicro.2014.12.026] [PMID: 25613122]

CHAPTER 8

Botton-up and Top-down Strategies for Processing Meat Analogs

Abstract: There is a worldwide concern regarding the supply of protein source for both human food and animal feed. In this context, meat has a major impact on the environment and consumers and health professionals have been expressing growing concern over high meat consumption. In this respect, meat analogue products - defined as compounds structurally similar to meat, however, differing in composition - have been attracting ever-increasing attention and interest of the environment and health concerned consumers. Furthermore, the number of people identifying as vegetarian and vegan have been increasing as they look for more sustainable options due to a unique lifestyle. In this chapter, the main aspects of bottom-up and top-down strategies are presented and discussed. The literature describes different approaches to simulate animal meat, such as cell culturing, mycoprotein and spinning techniques (bottom-up strategies) extrusion, proteins and metal-cation-precipitated polysaccharides, freeze structuring and shear cells technology (top-down strategies). Meat analogues have to present typical sensory and nutritional properties of animal meat for both vegetarian and non-vegetarian consumers. Thus, food scientists ought to test texturization techniques, develop more industrialized meat analogue products, evaluate the scalability of production process and quality control of meat analogues, improving the existing technologies and developing new ones that meet consumer demands.

Keywords: Bottom-Up Strategy, Cell Culturing, Extrusion, Polysaccharides, Protein, Shear Cells Technology, Top-Down Strategy.

INTRODUCTION

Since ancient times and up to this day, meat-consuming individuals are referred to as strong and powerful, making meat consumption a sign of social status [1]. Moreover, meat possesses a high biological protein value, due to its amino acid composition, vitamins and minerals [2]. However, its production requires a large number of natural resources, such as water, causing a large environmental impact [3, 4]. This fact contributes to the increased advocacy of plant proteins and their health benefits, justifying the importance of research and development of meat analogues [5].

Daneysa L. Kalschne, Marinês P. Corso & Cristiane Canan

In this scenario, meat analogues have been emerging as replacers [6], especially based on consumers' interest in plant-based products that could replicate meat in terms of mouthfeel, texture, taste, color and smell [4].

Moreover, animal cells and mycoproteins have also been studied as meat analogues for human consumption. Although meat analogues might replace meat in terms of its functionality since they achieve similar product properties and sensory attributes to meat [3], amino acids' biological value is also required. Meat analogues could be categorized as ground, comminuted and whole muscle. Fig. (**1**) shows some examples of meat analogue products, visually similar to animal meat.

Analogue
Hamburguer

Analogue
Meat ball

Analogue
Ground
beef

Fig. (1). Meat analogue products.

Different structuring techniques aimed at creating fibrous plant protein materials could be used to develop meat analogues. Dekker at al. classified these techniques as bottom-up and top-down strategies based on their key approach to create animal-meat-muscle-like fibrous structure [3]. According to the authors, the bottom-up strategy consists of creating anisotropic structural elements that are later assembled into larger products, whereas, the top-down strategy creates fibrous products by structuring biopolymer blends, in which the fibrousness of the product becomes apparent when stretching the material, mimicking the structure on larger length scales only. This nomenclature is used in the current chapter.

Plant protein sources, such as wheat, soy, pea, and peanut are the most widely used raw materials for the development of meat analogues [4 - 9]; however, some animal proteins, especially egg and milk-based [10, 11], and fungi protein [12] have also been studied. The literature describes several animal meat simulating techniques, such as cell culturing, mycoprotein and spinning (bottom-up strategy); extrusion, proteins and polysaccharides precipitated by metal cations, freeze structuring and shear cells technology (top-down strategy) [3]. However, high-moisture extrusion has emerged as a food processing technology applied for meat analogues production that has gained the interest of food scientists and is recognized as the most studied technique [5 - 9]. This chapter discusses some innovations concerning food structuring processes for protein texturization, allowing meat analogues production.

BOTTOM-UP STRATEGY

The bottom-up approach refers to a combination of individual structural elements to create subsequently assembled larger products [3]. It should be noted that meat is formed by the muscle (*in vivo*) composed of elongated cells arranged in parallel to form muscle fibers ranging from 10 to 100 μm in width and of a few mm in length [13]. Thus, the bottom-up strategy is associated with the combination of protein structures that could be grown by satellite-cell-culturing, filamentous fungi biomass production, or by creating protein fibrils or fibers from either a vegetal or animal protein source, depending on the adopted strategy.

CULTURED MEAT

Skeletal muscle-tissue engineering could be applied to cell-based meat production for human consumption [14]. To culture muscle fibers, initially, myoblast-form satellite cells should be harvested from the skeletal muscle of an animal; afterwards, the cells are replicated using a serum-supplemented medium with the nutrients needed for cell growth, such as amino acids, lipids, vitamins, and salts [3]. Cells culture media could be completely synthetic and devoid of serum products, however, serum-based media are still superior to synthetic ones [15].

After obtaining enough cells, they are placed onto a scaffold with anchor points to connect and align the cells, yielding a multicellular tissue. Cells alignment and muscle fibers development are ensured by either electric fields or other means, and after approximately three weeks, the muscle fibers would mature and could be harvested [3].

Satellite cells are undifferentiated and mononucleated [16], and according to Post [15], muscle cells culturing from satellite cells could be separated into two phases. The proliferation phase aims to obtain the highest number of cells from the cells of the starting batch, to maximize the number of doublings. Subsequently, the differentiation into skeletal muscle cells and maximum protein production (hypertrophy) occur. The author reinforces that differentiation in satellite cells occurs almost naturally with a very little adjustment to culture conditions. The cells merge forming myotubes, and start to express early-stage-skeletal-muscle markers. Hypertrophy occurs due to metabolic, biochemical and mechanical stimuli and it is suggested that mechanical stimuli are extremely important in triggering protein synthesis and protein organization into contractile units generating muscle typical striated microscopic morphology. Furthermore, cells in tissue engineering construction are cast in a collagen-like gel or onto a temporary biodegradable scaffold to simulate tendons formation.

A study on bovine satellite cell culture using Petri dishes was carried out by some researchers, and some of the articles can be accessed for further information [17 - 19]. For cells culturing upscaling, a large-scale reactor is needed for the myoblasts multiplication under the appropriate environmental stimuli. Mammalian cell sensitivity requires highly-controlled growing conditions in the reactor and contamination needs to be avoided [3].

The use of cell culture proteins isolated or mixed with plant-based meat analogues ought to be evaluated for consumer acceptance. Despite doubts about consumer responses regarding such meat products, Dutch research carried out on some of their population indicated that 63% of the respondents were in favor of the cultured meat concept and 52% were willing to try it [20], which justifies new researches on cell culturing meat.

MYCOPROTEIN

Mycoprotein is obtained from filamentous fungus *Fusarium venenatum* multiplication and it is used as a meat analogue and marketed as Quorn [12, 21]. *Fusarium venenatum* is multiplied using a continuous fermentation bioreactor in a glucose substrate under critical temperature and pH, which ought to be monitored and strictly controlled. After fermentation, the biomass is heated to obtain the RNA degradation and diffusion out of the cells. The residual biomass is heated and centrifuged obtaining a paste-like product with 20% solids (w/w) [3, 22 - 24]. The mycoprotein presents high protein and fiber contents and low fat, cholesterol, sodium, and sugar contents. Mycoprotein has been associated with the maintenance of healthy blood cholesterol levels, fostering of muscle synthesis, control of glucose and insulin levels and an increase in satiety [2, 12].

On the other hand, mycoprotein has also been associated with allergy and gastrointestinal events. Although the sensitization risk for Quorn is considered low (approximately 1/140,000 consumers), patients who are allergic to mould might react adversely to either inhaled or ingested mycoprotein [24, 25]. Symptoms such as pruritus, fainted/nearly fainted, urticaria (rash), anaphylaxis, periorbital, tongue and neck angioedema, with wheeze and shortness of breath, breathing difficulty, throat tightness, and vomiting [12, 24, 26] might occur.

Commercially available mycoprotein-produced products such as chunks, sausages, burgers, fillets, and steaks are usually prepared using a combination of mycoprotein and egg albumin (binding agent) [22].

SPINNING

Spinning is a traditional method used for producing anisotropic protein such as meat analogues from plant or animal protein sources. It induces biopolymer solutions alignment based on deformation and solidification [27].

Protein solution is extruded through a spinneret and subsequently bath-immersed containing a non-solvent that induces protein solidification [3, 28]. When a viscous biopolymer solution passes through a spinneret, the water-in-water emulsion is aligned and the resulting fibers are stretched. To obtain solidification, coagulation agents (salts, acid or alkali solutions) are used followed by washing of spun fibers [27]. The obtained fibers have the same thickness of the spinneret holes, usually in the order of 20 μm [3, 27].

The success of using spinning for protein texturization depends on protein source characteristics, such as denaturation conditions. For instance, when casein is used as a protein source, the fibers tend to melt together and lose their fibrous structure when heated in a moist environment forming a smooth plastic or molten mass. Thus the addition of heat-denaturable proteins (such as egg white or whey protein) could overcome this disadvantage [28]. Moreover, spinning involves some disadvantages, such as generating large waste streams from the coagulation and washing baths, the use of chemical coagulants, difficulty in ensuring microbiological safety due to high moisture content, and difficulty in assembling of single fibers into a food product on an industrial scale [27].

TOP-DOWN STRATEGY

The top-down approach refers to the structuring of biopolymer blends using an overall force field. These strategies include different mechanisms, such as a force field to direct biopolymer structure formation or aggregation, which would result in an anisotropic structure, ice crystals directional growth, fibrous products precipitation by metal cations, and shear force to create anisotropy. Even though the meat analogue structures obtained with these techniques mimic the animal meat structures, they differ from them in hierarchical architecture terms [3].

EXTRUSION

In order to obtain a larger market share, meat analogues produced with plant protein should be similar to animal meat rich in fibrous structure [5]. Extrusion is a traditional process used to produce anisotropic protein such as meat and seafood analogues from plant-sourced-protein [27]. In this process, a single protein or a protein blend is subjected to intensive heating and mixing in the screw barrel, and it is operated at high moisture levels and with a long cooling die attached to the end of the extruder, and the fibrous structure formation occurs only where the melted mixture is cooled and subsequently sheared [4, 6]. Fig. (2) presents an extrusion machine and fiber production from a protein scheme.

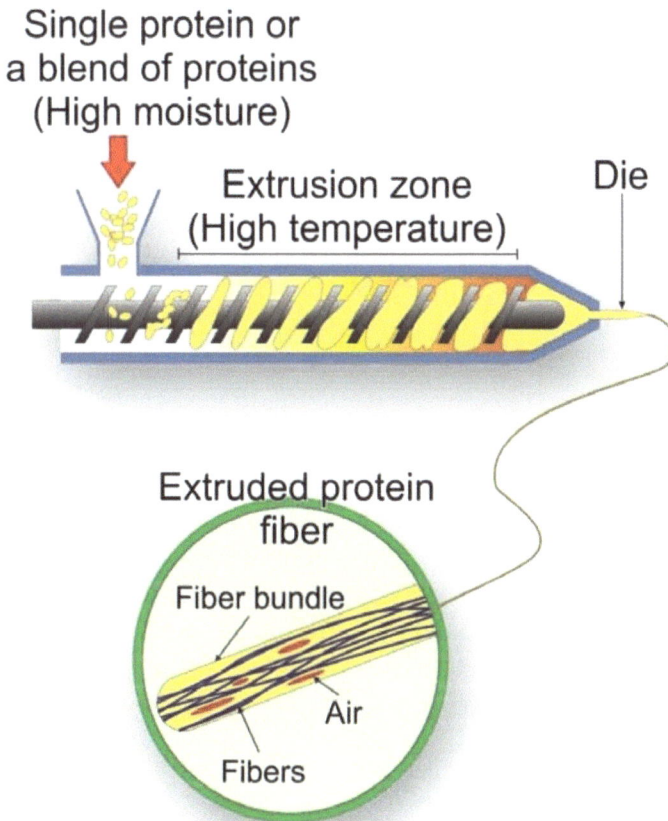

Fig. (2). Scheme of meat analogues fiber production by protein extrusion.

The production of palatable meat analogues using high moisture extrusion cooking depends on both the protein ingredients properties and the extrusion conditions [7]. Extrusion is carried out under high temperature, pressure and shear force [5, 29, 30] which change the structural and chemical properties of the protein. Parameters such as screw temperature, screw diameter, screw speed, screw configuration torque, moisture content, and feed rate should be controlled during extrusion [5, 6, 8].

High-moisture extrusion (moisture contents > 40%) offers advantages such as low energy input, no waste discharge, high efficiency, and higher texturized product quality due to rearrangements in protein structure [31]. Plant proteins could be texturized like animal meat, with rich fibrous structure and good springiness. The use of high-moisture extrusion technology has been deemed one of the best options to provide animal-meat-like texture with a rich fibrous and dense structure, strong springiness, and high moisture content [5].

The meat analogues production with the extrusion process is based on the fact that biopolymers form two or more separate phases, which should have high viscosity to allow deformation and alignment upon shearing, preserving the morphology after extrusion [32].

Peanut protein when underwent high-moisture extrusion (40%) to obtain a meat analogue, the authors confirmed the influence of extrusion parameters such as moisture on texturized peanut protein [5]. Moreover, pea protein was successfully texturized to highly fibrous extrudate at 55% moisture level, producing highly valuable protein ingredients for whole-muscle meat analogues production [7]. Furthermore, isolated soy, mung bean, peanut, pea, and wheat gluten proteins were texturized using 50% moisture extrusion for product characteristics comparison, and the results revealed that isolate pea protein presented the highest quality for meat analogues [8]. In addition, the use of lupin protein, *Spirulina* biomass and iota carrageenan along with high moisture extrusion was suitable for obtaining meat analogues with enhanced physical properties [29]. Similarly, 1.5% iota-carrageenan addition showed a positive impact on the moisture, texture and sensory parameters of a soy protein meat analogue prepared with high moisture extrusion process [33]. Meat analogues were also produced by the high moisture extrusion of wheat gluten [6]. However, gluten has been also associated with health disorders such as celiac disease, wheat allergy, and non-celiac gluten sensitivity [34]. A number of plant proteins possess such an allergenic aspect, which ought not to be overlooked when dealing with consumers allergic to the foods studied.

Protein extrusion causes dramatic structural changes to protein molecules and unfolding in the extruder barrel, creating favorable molecular rearrangement conditions in the subsequent extrusion zones, drawing the fibrous structure near to the animal meat [31]. Air inclusion is of utmost importance in dense protein materials used for fibrous material production [32].

FREEZE STRUCTURING

The freeze structuring (freeze alignment) is the least used and studied technology. It ought to be applied to a protein soluble in a frozen aqueous solution in order to lead to an anisotropic structure. However, if heat is removed unidirectionally without mixing, the ice crystal needles alignment will yield anisotropic structures. After that, the frozen block is dried by freeze-drying to obtain a porous microstructure with a sheet-like parallel protein orientation [3]. The structure obtained is insoluble as opposed to the initial protein.

PROTEINS AND HYDROCOLLOIDS PRECIPITATED BY METALS CATIONS

A patent described the use of a protein (curd, cheese, powdered milk or similar) and a hydrocolloid (pectin with a low methoxyl group content, gellan gum and alginate) precipitated by metal cations (valency 2) in water to produce a meat analogue. A high temperature was used for the homogeneous mixture formation, obtaining a fibrous product [11]. The patent also reported that the fibrous product was rinsed to remove any metal cations residue and pressing was used to remove any excess water, obtaining a 40 to 60% (w/w) dry matter content.

According to Dekkers *et al.* , plant proteins such as soy, rice, maize and lupine could be used in a similar way to that of milk proteins for meat analogues production. The authors also pointed out that the process is scalable, yielding products with some degree of structure, yet relatively using intense resource [3]. Meat analogues from milk proteins have been available in the Dutch market since 2005 under the brand name "Valess", followed by Belgium in 2007 and Germany in 2009. The available products were breaded and un-breaded filled filets for cooking and frying.

SHEAR CELL TECHNOLOGY

Shear cell technology combines shear flow and heat in a process that introduces fibrous structural patterns to a plant protein granular mixture [4]. The shear cell is a cone–cone device based on a cone–plate rheometer, where, the first cone is stationary, while the bottom one rotates. Both cones are heated and cooled aided by an oil bath, and as opposed to extrusion, deformation occurs inside the device as a constant process, and the simple shear and heat combination aligns the proteins forming fibrous structures [30].

A pre-heated shear cell at high moisture (60%), with temperatures ranging from 95 to 140 °C and kept at constant shearing and heating for 15 min was applied to pea protein blended with wheat gluten. At 120 °C, the mixture resulted in elongated protein domains and air bubbles along the shear flow direction, which appeared as aligned fibers in the product. In contrast, a lower temperature processing resulted in a weak product without any fibers and higher temperatures resulted in a decrease of strength [32]. In addition, the study has also reported that pea protein blended with wheat gluten offers opportunities for fibrous meat analogues formation with matrix strength similar to that of cooked chicken. Also, some studies described the use of a lab-scaled Couette Cell to modify a soy and gluten protein mixture. The results showed that it is possible to obtain highly anisotropic fibrous samples to be used as meat analogues by simple shear flow and heat (120 °C, 30 min and 20 RPM) [4, 30].

FINAL CONSIDERATIONS

The technologies for meat analogues production presented in this chapter were developed some time ago. However, in recent years, consumer perceptions of animal food production and consumption have changed and a large number of consumers have moved to either vegetarian or vegan diets. Besides the non-vegetarian consumers, meat is a well-appreciated food worldwide illustrating social status. To attract both vegetarian and non-vegetarian consumers, meat analogues have to present typical sensory properties of animal meat, such as aroma, texture, flavor, juiciness, and chewiness. Consumers also expect to find in meat analogues the same nutritional properties found in animal meat. The combination of two or more texturization techniques and the development of a greater number of industrialized meat analogue products ought to be tested due to the scarcity of applied studies. The importance of scalability of meat analogues productive process and quality control ought to be also considered. The path ahead is a long one for food scientists conducting researches on meat analogues as well as improving existing technologies and developing new ones to meet consumer demands.

REFERENCES

[1]   E.Y. Chan, and N. Zlatevska, "Jerkies, tacos, and burgers : Subjective socioeconomic status and meat preference", *Appetite,* vol. 132, pp. 257-266, 2019.
        [http://dx.doi.org/10.1016/j.appet.2018.08.027]

[2]   V. Joshi, and S. Kumar, "Meat Analogues: Plant based alternatives to meat products- A review", *Int. J. Food Ferment. Technol.,* vol. 5, no. 2, p. 107, 2015.
        [http://dx.doi.org/10.5958/2277-9396.2016.00001.5]

[3]   B.L. Dekkers, R.M. Boom, and A.J. Van Der Goot, "Trends in food science & technology structuring processes for meat analogues", *Trends Food Sci. Technol,* vol. 81, pp. 25-36, 2018.
        [http://dx.doi.org/10.1016/j.tifs.2018.08.011]

[4]   G.A. Krintiras, J. Gadea Diaz, A.J. Van Der Goot, A.I. Stankiewicz, and G.D. Stefanidis, "On the use of the Couette Cell technology for large scale production of textured soy-based meat replacers", *J. Food Eng.,* vol. 169, pp. 205-213, 2016.
        [http://dx.doi.org/10.1016/j.jfoodeng.2015.08.021]

[5]   J. Zhang, L. Liu, Y. Jiang, S. Faisal, and Q. Wang, "A new insight into the high-moisture extrusion process of peanut protein : From the aspect of the orders and amount of energy input", *J. Food Eng.,* vol. 264, p. 109668, 2020.
[http://dx.doi.org/10.1016/j.jfoodeng.2019.07.015]

[6]   V.L. Pietsch, R. Werner, H.P. Karbstein, and M.A. Emin, "High moisture extrusion of wheat gluten : Relationship between process parameters, protein polymerization, and final product characteristics", *J. Food Eng.,* vol. 259, pp. 3-11, 2019.
[http://dx.doi.org/10.1016/j.jfoodeng.2019.04.006]

[7]   R. Osen, S. Toelstede, F. Wild, P. Eisner, and U. Schweiggert-weisz, "High moisture extrusion cooking of pea protein isolates : Raw material characteristics, extruder responses, and texture properties", *J. Food Eng.,* vol. 127, pp. 67-74, 2014.
[http://dx.doi.org/10.1016/j.jfoodeng.2013.11.023]

[8]   S. Samard, and S. Samard, "Physicochemical and functional characteristics of plant protein-based meat analogs", *J. Food Process. Preserv.,* pp. 1-11, 2019.
[http://dx.doi.org/10.1111/jfpp.14123]

[9]   S.M. Beck, K. Knoerzer, and J. Arcot, "Effect of low moisture extrusion on a pea protein isolate's expansion, solubility, molecular weight distribution and secondary structure as determined by Fourier Transform Infrared Spectroscopy (FTIR)", *J. Food Eng.,* vol. 214, pp. 166-174, 2017.
[http://dx.doi.org/10.1016/j.jfoodeng.2017.06.037]

[10]   N. Akramzadeh, H. Hosseini, Z. Pilevar, N.K. Khosroshahi, K. Khosravi-Darani, R. Komeyli, F. J. Barba, A. Pugliese, M.M Poojary, and A.M. Khaneghah, "Physicochemical properties of novel non☐meat sausages containing natural colorants and preservatives", *J. Food Process. Preserv.,* vol. 42, no. 9, pp. 1-8, 2018.
[http://dx.doi.org/10.1111/jfpp.13660]

[11]   A. C. Kweldam, G. Kweldam, and A. C. Kweldam, *Method for the preparation of a meat substitute product, meat substitute product obtained with the method and ready to consume meat substitute product,* US20110244090A1, 2011.

[12]   T.J.A. Finnigan, B.T. Wall, P.J. Wilde, F.B. Stephens, S.L. Taylor, and M.R. Freedman, "Mycoprotein: the future of nutritious nonmeat protein, a symposium review", *Curr. Dev. Nutr.,* vol. 3, no. 6, pp. 1-5, 2019.
[http://dx.doi.org/10.1093/cdn/nzz021] [PMID: 31187084]

[13]   G. Strasburg, Y. Xiong, and W. Chiang, "Fisiologia e Química dos tecidos musculares comestíveis", In: *Química de Alimentos de Fennema.,* S. Damodaran, K.L. Parkin, O.R. Fennema, Eds., Artmed: Porto Alegre, 2010, pp. 719-757.

[14]   R. Simsa, J. Yuen, A. Stout, N. Rubio, P. Fogelstrand, and D.L. Kaplan, "Extracellular heme proteins influence bovine myosatellite cell proliferation and the color of cell-based meat", *Foods,* vol. 8, no. 10, 2019.
[http://dx.doi.org/10.3390/foods8100521] [PMID: 31640291]

[15]   M.J. Post, "Cultured meat from stem cells: challenges and prospects", *Meat Sci.,* vol. 92, no. 3, pp. 297-301, 2012.
[http://dx.doi.org/10.1016/j.meatsci.2012.04.008] [PMID: 22543115]

[16]   R.M.S.A. Foschini, F.S. Ramalho, and H.E.A. Bicas, "Células satélites musculares", *Arq. Bras. Oftalmol.,* vol. 67, no. 4, pp. 681-687, 2004.
[http://dx.doi.org/10.1590/S0004-27492004000400023]

[17]   K.J. Thornton, E. Kamanga-Sollo, M.E. White, and W.R. Dayton, "Active G protein-coupled receptors (GPCR), matrix metalloproteinases 2/9 (MMP2/9), heparin-binding epidermal growth factor (hbEGF), epidermal growth factor receptor (EGFR), erbB2, and insulin-like growth factor 1 receptor (IGF-1R) are necessary for trenbolone acetate-induced alterations in protein turnover rate of fused bovine satellite cell cultures", *J. Anim. Sci.,* vol. 94, no. 6, pp. 2332-2343, 2016.

[http://dx.doi.org/10.2527/jas.2015-0178] [PMID: 27285910]

[18]   E. Kamanga-Sollo, M.S. Pampusch, G. Xi, M.E. White, M.R. Hathaway, and W.R. Dayton, "IGF-I mRNA levels in bovine satellite cell cultures: effects of fusion and anabolic steroid treatment", *J. Cell. Physiol.*, vol. 201, no. 2, pp. 181-189, 2004.
[http://dx.doi.org/10.1002/jcp.20000] [PMID: 15334653]

[19]   E. Kamanga-Sollo, M.E. White, W.J. Weber, and W.R. Dayton, "Role of estrogen receptor-α (ESR1) and the type 1 insulin-like growth factor receptor (IGFR1) in estradiol-stimulated proliferation of cultured bovine satellite cells", *Domest. Anim. Endocrinol.*, vol. 44, no. 1, pp. 36-45, 2013.
[http://dx.doi.org/10.1016/j.domaniend.2012.08.002] [PMID: 23036864]

[20]   M.J. Post, "Cultured beef: medical technology to produce food", *J. Sci. Food Agric.*, vol. 94, no. 6, pp. 1039-1041, 2014.
[http://dx.doi.org/10.1002/jsfa.6474] [PMID: 24214798]

[21]   J. Lonchamp, P. S. Clegg, and S. R. Euston, "Foaming, emulsifying and rheological properties of extracts from a co-product of the Quorn fermentation process", *Eur. Food Res. Technol.*, no. 0123456789, 2019.
[http://dx.doi.org/10.1007/s00217-019-03287-z]

[22]   M.G. Wiebe, "Myco-protein from Fusarium venenatum: a well-established product for human consumption", *Appl. Microbiol. Biotechnol.*, vol. 58, no. 4, pp. 421-427, 2002.
[http://dx.doi.org/10.1007/s00253-002-0931-x] [PMID: 11954786]

[23]   M.G. Wiebe, "QuornTM myco-protein - Overview of a successful fungal product", *Mycologist*, vol. 18, no. 1, pp. 17-20, 2004.
[http://dx.doi.org/10.1017/S0269915X04001089]

[24]   N. Maina, and M. Sikora, "P330 A case report of possible immediate hypersensitivity to ingested quorn in a mold allergic patient", *Ann. Allergy Asthma Immunol.*, vol. 119, no. 5, p. S78, 2017.
[http://dx.doi.org/10.1016/j.anai.2017.08.221]

[25]   S.J. Katona, and E.R. Kaminski, "Sensitivity to Quorn mycoprotein (Fusarium venenatum) in a mould allergic patient", *J. Clin. Pathol.*, vol. 55, no. 11, pp. 876-877, 2002.
[http://dx.doi.org/10.1136/jcp.55.11.876-a] [PMID: 12401831]

[26]   M.F. Jacobson, and J. DePorter, "Self-reported adverse reactions associated with mycoprotein (Quorn-brand) containing foods", *Ann. Allergy Asthma Immunol.*, vol. 120, no. 6, pp. 626-630, 2018.
[http://dx.doi.org/10.1016/j.anai.2018.03.020] [PMID: 29567357]

[27]   J.M. Manski, A.J. van der Goot, and R.M. Boom, "Advances in structure formation of anisotropic protein-rich foods through novel processing concepts", *Trends Food Sci. Technol.*, vol. 18, no. 11, pp. 546-557, 2007.
[http://dx.doi.org/10.1016/j.tifs.2007.05.002]

[28]   C.R. Southward, "Casein and Caseinates. Uses in the Food Industry", In: *Encyclopedia of Food Sciences and Nutrition.*, B. Caballero, P.M. Finglas, F. Toldra, Eds., vol. 1. 2nd ed. Academic Press, 2003, pp. 948-958.
[http://dx.doi.org/10.1016/B0-12-227055-X/00179-6]

[29]   M. Palanisamy, S. Töpfl, R. G. Berger, and C. Hertel, "Physico-chemical and nutritional properties of meat analogues based on Spirulina/lupin protein mixtures", *Eur. Food Res. Technol,* no. 0123456789, 2019.
[http://dx.doi.org/10.1007/s00217-019-03298-w]

[30]   G.A. Krintiras, J. Göbel, A.J. Van Der Goot, and G.D. Stefanidis, "Production of structured soy-based meat analogues using simple shear and heat in a Couette Cell", *J. Food Eng.*, vol. 160, pp. 34-41, 2015.
[http://dx.doi.org/10.1016/j.jfoodeng.2015.02.015]

[31]   J. Zhang, L. Liu, Y. Jiang, S. Faisal, and Q. Wang, "Food and beverage chemistry / biochemistry

converting peanut protein biomass waste into ' double green ' meat substitutes using a high-moisture extrusion process : A multi-scale method to explore a process for forming a meat-like fibrous structure convert", *J. Agric. Food Chem,* vol. 67, no. 38, pp. 10713-10725.
[http://dx.doi.org/10.1002/jsfa.6474]

[32]   F.K.G. Schreuders, "Comparing structuring potential of pea and soy protein with gluten for meat analogue preparation", *J. Food Eng.,* vol. 261, pp. 32-39, 2019.
[http://dx.doi.org/10.1016/j.jfoodeng.2019.04.022]

[33]   M. Palanisamy, S. Töpfl, K. Aganovic, and R.G. Berger, "Influence of iota carrageenan addition on the properties of soya protein meat analogues", *Lebensm. Wiss. Technol.,* vol. 87, pp. 546-552, 2018.
[http://dx.doi.org/10.1016/j.lwt.2017.09.029]

[34]   P.V.G. Resende, N. L. de M. e Silva, G. C. M. Schettino, and P. M. F. Liu, "Gluten related disorders", *Rev. Médica Minas Gerais,* no. Suplementar 3, pp. S51-S58, 2017.
[http://dx.doi.org/10.5935/2238-3182.20170]

CHAPTER 9

Alternative Additives and Ingredients

Abstract: Meat products conventionally contain fat, saturated fatty acids, and high salt contents, as well as health-harming chemical additives, such as nitrites and synthetic anti-oxidants. The meat industry has been striving to find alternatives in order to meet consumer demands for healthier products, either reduced or free of such components. For this purpose, various studies have been carried out for alternative ingredients. Vegetable oils have achieved positive results as fat replacers. Methods such as pre-emulsion, emulsion-templated, microencapsulation, and oleogel formation (hydrogel or organogel) have been excellent on partial pork fat reduction, and consequently saturated fatty acids or n-3 PUFA-rich oils incorporation. Plant materials are seen as a good alternative for synthetic anti-oxidants. Various plant derivatives have been tested and presented anti-oxidant potential due to their bioactive contents, such as phenolic acids, phenolic diterpenes, flavonoids, and volatile oils. Other salts, especially potassium chloride, have stood out on replacing sodium chloride. The use of natural sources of either nitrites or nitrates, such as celery and powdered vegetable juice, has been suggested as their replacers, as well as the use of different compounds with potential natural food preservative characteristics. Overall, considerable progress has been made over the past few years in the field of non-meat ingredients as alternative to conventional ones, conferring a healthier approach to meat products. However, overall strategic approaches to meet one or more simultaneous demands, such as fat and salt reduction, among others, in terms of sensory, technological and microbiological efficiency ought to be encouraged.

Keywords: Animal Fat Replacer, Natural Anti-Oxidants, Nitrite-Free Product, n-3 PUFA, Potassium Chloride, Salt Replacer.

INTRODUCTION

Consumer demand for healthier meat products has increased in recent years following overall diet changes. To meet this demand, the meat industry has been striving to innovate on the processing of healthier products, either by reducing the number of compounds with negative health impacts or by incorporating ingredients that could offer benefits to consumer health when included in the diet.

Consumer evaluations for reformulated and healthier processed meat products showed that healthy ingredients, salt, and/or fat contents are the most important factors on consumers' purchase intention, second only to price and meat-based

Daneysa L. Kalschne, Marinês P. Corso & Cristiane Canan

products. For reformulation, consumers preferred salt and/or fat-reduced products over non-reduced ones. For healthy ingredients, Omega 3 was the preferred one and healthier reformulations improved the health perception of processed meats [1].

Typically, meat products belong to a caloric food category due to their high-fat content, mainly saturated fat, high sodium content and chemical additives such as nitrates and nitrites, synthetic anti-oxidants, polyphosphates, or even allergenic compounds, such as soy proteins, milk proteins and others that are associated with health problems. Such ingredients have been widely used by the meat industry due to their technological functions, sensory properties, microbiological safety, and the economic viability that they confer to meat products.

Therefore, this chapter aims at addressing alternative replacements to the negative connotations of more conventional ingredients and additives (fat and/or saturated fatty acid, synthetic anti-oxidants, salt, nitrates, and nitrites), aiming at a healthier approach to meat products.

ANIMAL FAT REPLACERS

The high animal fat contents found in meat products are comprised of high saturated fatty acid content. Pork fatback is the most commonly used fat, providing the product with the taste and texture characteristics that ensure its sensory acceptance by consumers. However, consumers aim for healthy products as well as sensory quality and the meat industry has been looking for pork fat replacers, given the difficulty of bridging sensory quality, low saturated fatty acids concentration, n−3 PUFA-rich oils, and oxidative stability of products.

Given the inefficiency of applying vegetable oils directly to the products, more satisfactory results have been verified from the application of pre-emulsified vegetable oil, as previously covered in chapter 3. Oil pre-emulsification increases protein-covered fat globules with consequently greater batter stability. Kang *et al.*, using pre-emulsified soy oils to replace pork fatback changed the protein structure (higher ß-sheet and lower α-helix content, tyrosine residues were exposed, and more hydrophobic interactions were formed) allowing for lower fat and energy content, enhancing the quality of frankfurters [2]. Utama *et al.*, when using oil-i--water (o/w) emulsion, consisting of perilla and canola oil, polyglycerol polyricinoleate, soy protein isolate, and inulin as an animal fat replacer in chicken sausage found that the emulsion could effectively improve the fatty acid profile and oxidative stability without deteriorating the sensory characteristics of the emulsified sausage [3]. Hu *et al.*, using pre-emulsified soybean oil with regenerated cellulose and sodium caseinate in emulsified sausages [4], and Pintado *et al.*, using water and chia flour with pre-emulsified olive oil in frankfurters have also achieved satisfactory outcomes [5].

Another approach for pork fat replacement is the incorporation of vegetable oils in oleogel-form (gel in that the liquid phase if oil). The vegetable oil is structured by a gelling compound that, by heating and cooling, is stacked in organized three-dimensional networks, thus providing the vegetable oil nutritional benefit and the positive sensory and technological properties of a more saturated and harder animal fat [6]. Several gelling compounds have been suggested, including kappa-carrageenan [7, 8], hydroxypropyl methylcellulose [9], monoglycerides and phytosterols [10], rice bran wax [6], ethylcellulose [8, 11], alginate [12], and others. According to Taverniers *et al.*, polymers appear to be the most promising candidates since a considerable number have been approved for use in food products, capable of providing structure, as well as being widely available and cost-effective [13]. The most recent oleogel application in meat products and the main results are presented in Table 1.

Table 1. Examples of oleogel applied in meat products as pork fat replacers

Meat product	Oleogel composition	Mainly results	Reference
Meat batters	Hydrogel: Canola oil (40%), polysorbate 80 (oil: surfactant ratio was 0.05%, 0.003), BHT (0.01%), kappa carrageenan (1.5% or 3%), and deionized water (up to 100%) Organogel: Canola oil (88%, 86.5% or 85%), ethylcellulose (12%), glycerol monostearate (0%, 1.5% or 3%), BHT (0.01%).	Replacement by regular canola oil increased hardness and lightness when compared to those with beef fat Batters formulated with organogel showed enhanced matrix stability compared to those with hydrogelled emulsions.	[8]
Meat patties	Hydrogel: Hydroxypropyl methylcellulose dissolved in distilled water (1%, w/w) overnight and added with canola oil (2, 4, and 6%, w/w).	The 50% replacement of beef tallow by HPMC oleogels reduced cooking loss, improved saturated to unsaturated fat ratio from 0.73 to 0.18 and produced meat patties with good sensory properties.	[9]
Frankfurter-type sausages	Rice bran wax (2.5% and 10% wt/wt) and correct amounts of soybean oil.	Oleogel treatments were less dark and less red with reduced flavor. It was similar to pork fat treatment for firmness, chewiness, springiness and aroma.	[6]
Frankfurter sausages	Sunflower oil (80%), gelators' (20%) (monoglycerides/phytosterols weight ratios were 10:10 (1:1) and 15:5 (3:1) w/w).	For the control, no differences were detected in the oxidation levels and sensory evaluation.	[10]

(Table 1) cont.....

Meat product	Oleogel composition	Mainly results	Reference
Frankfurter sausages	Organogel: Ethylcellulose (8, 10, 12, and 14%), sorbitan monostearate (0, 1,5 and 3%) and canola oil (correct amounts).	The use of organogel resulted in a hardness value similar to that of beef fat, by both sensory and texture profile analyses, but with a lighter color. Smokehouse yields were higher for canola oil- and organogel-containing treatments.	[11]

According to Taverniers *et al.* [13], the emulsion-templated method is easily applied. Various food polymers such as proteins and modified polysaccharides are capable of stabilizing oil-in-water emulsions due to their amphiphilic property. After the adsorption of such polymers to the oil-water interface and the stable emulsions formation, water could be removed until the system contains more than 95% of oil by weight, forming a physical network of polymers that trap the liquid oil.

The microencapsulated n-3 PUFA-rich oils have also been used with quite promising results. Microencapsulation increases n-3 PUFA levels resistant to pH and temperature during processing without affecting the hardness, enhancing important technological properties such as cooking loss, fat retention and healthier PUFA / SFA and n − 6 / n − 3 ratios. This technique was successfully applied in the incorporation of chia and linseed oils encapsulated with sodium alginate in hamburgers [14] and fish oil encapsulated with filled hydrogel particles in frankfurters and meat system [15, 16].

Another approach for animal fat replacement, besides the use of vegetable oils, is fat reduction by incorporating soluble- and insoluble-rich fiber ingredients. The addition of fiber in meat products improves both texture and yielding properties, as well as their nutritional value. Some studies have also investigated the use of fiber combined with pork or poultry skin, which due to their high collagen contents, favors texture and yield properties, as well as costs reduction. Several studies with different fibers or fibrous materials and their combination with skin have reported significant fat reductions in meat products without harming the technological and sensory properties. Choe *et al.*, using pork skin and wheat fiber mixture reduced up to 50% pork fat in frankfurters [17]. Alves *et al.*, using a mixture of pork skin and green banana flour, reduced pork fat by up to 60% in Bologna-type sausages [18]. Choi *et al.* reduced approximately 33% of pork fat by incorporating makgeolli lees fiber in frankfurters [19]. Promising results were also obtained by Han *et al.* using insoluble fiber and soy hull-extracted pectin [20] and by Kim *et al.* using sugarcane bagasse fiber [21].

NATURAL ANTI-OXIDANTS

Meat and poultry processors have used synthetic anti-oxidants such as butylated hydroxyanisole (BHA), butylated hydroxytoluene (BHT), tert-butylhydroquinone (TBHQ), and propyl gallate, as well as tocopherols, to prevent lipid and protein oxidation [22]. Such anti-oxidants reinstate free radicals, thus delaying lipid oxidation and the development of unpleasant flavors, as well as improving color stability. The loss of sensory characteristics in meat products could lead to dissatisfied consumers and economic loss. In the food industry, anti-oxidants are divided into natural and synthetic. The use of synthetic anti-oxidants has been questioned due to their toxicological and carcinogenic effects. Therefore, the food industry has opted for natural products over synthetic ones to meet consumer demand [22 - 25].

Plant materials provide good alternatives for synthetic anti-oxidants. Spices and herbs, often used for their flavoring characteristics, could be added as whole, ground, or isolated from their extracts to meat products. These natural anti-oxidants contain some active compounds (phenolic acids, phenolic diterpenes, flavonoids, and volatile oils) exhibiting anti-oxidant potential in meat products. Each of these compounds possesses strong hydrogen donation activity to oxygen radicals, making them extremely effective anti-oxidants. Some compounds could also chelate metals, and thus slowing down the oxidation process by two mechanisms [26 - 28]. Given the high effectiveness of some bioactive compounds, the use of natural herbal anti-oxidants by the meat industry has been noted. Numerous studies on natural anti-oxidants in meat products have been found in the literature with promising results. Tomovic *et al.* [26] listed 52 manuscripts published from 2002 to 2017 on the use of plant materials in meat products such as ascorbic acid, α-tocopherol, sesamol, basil essential oil, black currant extract, black mustard (*Brassica nigra)*, Chinese cinnamon (*Cinnamomum cassia)*, oregano (*Origanum vulgare)* (extract and essential oil), clove (*Syzygium aromaticum),* butterbur extract, broccoli extract, caraway essential oil, pomegranate (rind powder extract, juice and seed powder extract), grape extracts (seed and peel), citrus extract, honey, du-zhong extracts (leaf, roasted cortex, and seed), *Echinacea angustifolia* extracts, green tea extract, pine bark extract, rosemary (oleoresin and extracts), fluoxetine (*Hypericum perforatum* L.) extract, juniper (*Juniperus communis* L.), mushroom extract, nutmeg, sage essential oils, avocado (peel, pulp, and seed), purple rice bran extract, rooibos (dried leaves, water extract and freeze-dried extract), lemon balm, coffee, garlic, spice extracts, tamarillo, thyme and balm essential oils, and winter savory essential oil. More recent studies have also reported phytic acid extracted from rice bran [29], aqueous and ethanolic tomato extracts [27, 28], olive aqueous extracts [27] powdered lotus leaves, shepherd's purse and golden rod [30], cumin and cardamom [31], green coffee bean extract [32], coffee residue extracts [33], beer residue extract and peanut skin extract [34], hull, bur and leaf chestnut (*Castanea sativa*) extracts [34, 35], jabuticaba (*Myrciaria cauliflora*) aqueous extract [36], pitanga (*Eugenia uniflora* L.) leaf extracts [37], and Guarana (*Paullinia cupana*) seed extracts [38].

According to Ribeiro *et al.* [39], the use of natural anti-oxidants in active packaging for meat products should be exploited since they reduce the use of potentially toxic synthetic anti-oxidants. The authors listed various studies on natural anti-oxidants in films applied to meat products, such as ascorbic acid, ferulic acid, quercetin, and green tea extract in ethylene-vinyl alcohol copolymer (EVOH); rosemary extract in low-density polyethylene film, or gelatin- or polyamide-based edible films; oregano extract in gelatin- or polyethylene-based edible films; oregano essential oil and green tea extract in PET/PE/EVOH/PE films; green tea, oolong tea (partially fermented) and black tea (fermented) (*Camellia sinensis*) extracts in protein films; barley husks in low-density polyethylene film; radish extract in microcapsule-coated ethylene-vinyl acetate films; and olive leaves *(Olea europaea L.)* extract in polyethylene - LDPE multilayer films.

As oxidation is the main cause of meat and meat products deterioration, the use of anti-oxidants is essential and alternatives to synthetic anti-oxidants have been sought to meet consumer demand. Despite the significant number of studies aimed at finding alternatives, there is still much to explore as anti-oxidant compounds could be found on the leaves, fruits, seeds, and bark of almost all plants. The anti-oxidant potential could be altered depending on the harvesting, storage, extraction, and application conditions. Also, the anti-oxidant effect will vary depending on the meat product, given the different compositions and processing steps to which each product is submitted.

Some of the studied natural anti-oxidants are condiments and spices traditionally used in meat products. According to Aminzare *et al.*, the use of such natural additives poses some limitations due to their intense and rich flavor, which might have adverse sensory effects on meat products. Further studies ought to investigate new technologies such as high-pressure processing, pulsed electric fields, and ultrasound combined with natural anti-oxidants to enhance the anti-oxidant properties of meat products [25].

SALT REPLACERS

Sodium chloride is the most commonly used additive in food processing. Salt confers flavor, shelf-life, and texture to meat products. Salt (normally 1.5 to 2.5% sodium chloride) enhances gelation, water binding capacity, fat retention, and cooking loss in meat batters. However, high salt intake is associated with negative effects on blood pressure and cardiovascular system. It urges the meat industry to look for alternatives, which include different approaches such as reducing salt addition by either replacing it with different metal salts, flavor enhancers, or even by applying technologies such as ultrasound and high-pressure processing combined with replacers for enhanced microbiological safety [40, 42].

Among the prospective additives to replace sodium chloride are included other chlorinated salts (KCl, $CaCl_2$, and $MgCl_2$) that pose no threat to the consumers' health. Potassium chloride (KCl) has been most studied as it has presented better outcomes than calcium chloride ($CaCl_2$) and magnesium chloride ($MgCl_2$) on the physicochemical and sensory quality of meat products [43, 44]. Despite that, studies showed the use of KCl might increase bitter, astringency, and metallic taste. This negative effect was minimized with the combined use of KCl, 1% lysine, and/or 0.1% liquid smoke to replace 50% NaCl [45]. Triki *et al.*, found positive outcomes for technological, sensory, and microbiological attributes using a 50% KCl, 35% $MgCl_2$, and 15% $CaCl_2$ salt mixture to replace up to 100% NaCl [46].

Pires *et al.* report the use of PuraQ® Na4 commercial product, especially developed to replace salt in cured and non-cured processed meat products as well as to control their water activity, reducing the microbial growth rate and concluded that 40% replacement of salt retained good sensory quality and delivered a healthier product [45]. Seganfredo *et al.* also applied PuraQ® Arome Na4 commercial product to replace up to 30% NaCl in Toscana sausage, obtaining good results in terms of texture and sensory acceptance [47]. Inguglia *et al.* listed several commercial products developed for this purpose with promising results. The authors also reported that a salt physical form change from granular to flaked, for instance, provided satisfactory sodium reduction results due to higher solubility. Examples of commercially available modified salt crystals are Alberger® Flake Salt and Star Flake® salt from Cargill [42].

Cox *et al.* [48] verified that sodium content could be potentially reduced in model reduced and low-fat oil-in-water emulsion systems, not being detected by the consumer. Having the insight for where consumers are able to detect differences in sodium contents could contribute to a successful sodium reduction.

Binding agents might be used to reduce salt content without impairing the solubilization function of myofibrillar proteins. These ingredients enhance meat pieces binding in restructured or reformed meat products and/or increase the water binding and emulsification capacity of the product. A wide range of ingredients, including functional proteins (soy or milk proteins), polyphosphates, fibers, hydrocolloids, starches, and microbial transglutaminase, have been explored [40 - 42, 49].

Overall, effective options to achieve the sensory and technological properties provided by salt have been observed in the literature and commercially. In terms of microbiological safety, advances have been made in the use of emerging technologies such as ultrasound and high-pressure combined with salt replacing ingredients. However, considering the health problems caused by high salt intake, strategic studies integrating technological sensory efficiency, microbiological safety, and cost-compatible salt ought to be encouraged to meet the demands of both the meat industry and consumers.

NITRITE AND NITRATE REPLACERS

Nitrite, just as salt, is a multifunctional ingredient used in cured meat. A small concentration (50-100 ppm) is enough to achieve important functions in meat products. Both sodium and potassium nitrite are responsible for the reactions with myoglobin, the stabilization of red cured meat color, the cured flavor, and the flavor protection due to its anti-oxidant action and bacterial inhibition, mainly the outgrowth and neurotoxin formation by *Clostridium botulinum* [50, 51]. However, nitrosating agents might react with secondary amines forming N-Nitrosamines. Several compounds found in meat and meat products could act as precursors to the carcinogenic N-nitrosamines formation [52]. Therefore, strategies to replace nitrite/nitrate in meat products have been encouraged in order to produce low-or nitrite/nitrate-free products by means of different alternatives for color, oxidation, and microbiology safety.

The use of ingredients that are natural nitrite and nitrate sources (reduced to nitrite by microorganisms, *e.g. Staphylococcus*), such as celery and vegetable juice powder, were indicated to avoid direct addition of nitrite to meat products [53, 56]. In some of these studies, a combination with ingredients such as sodium lactate, which possesses a conservative action, cochineal, and orange dietary fiber by colorant effects and vitamin C and E anti-oxidants showed positive outcomes on hot dog sausage production. Aquilani *et al.* [57], when replacing sodium nitrite by natural anti-oxidant mixtures using grape seed extract, olive pomace hydroxytyrosol, and chestnut extract in dry-fermented pork sausages reported effects on color.

García *et al.* reported that chitosan, a potential natural food preservative in meat products, could be used in pork sausage without impairing either color or flavor, increasing their shelf life. However, panelists detected textural differences among the samples with chitosan, suggesting potential influence of chitosan deacetylation degree on the sausages' textural behavior, requiring further investigation [58].

Canan *et al.* [59] verified the efficient antimicrobial action of a phytate-based composition (phytic acid salt commonly called phytin), sodium inositol hexaphosphate, or sodium myo-inositol hexaphosphate added with maltodextrin on *Clostridium perfringens*. Phytates are natural, stable substances, which allow for their application in different food matrices. This phytate-based ingredient might be applied in a solid or liquid form, pure, diluted, or with other additives, at varied dosages as per the microbiological and physicochemical characteristics of the *C. perfringens* contaminated food matrix and/or present conditions for its development.

The outcomes for nitrate and nitrite replacement are noteworthy but the need for combined strategies to achieve all functions of such additives must be emphasized in order not to impair the sensory acceptance (mainly color) and microbiological safety of the products.

CONCLUSION

Overall, considerable progress has been made over the last few years in the field of alternative non-meat ingredients as they could confer a healthier product approach to meat products. Despite such achievements, global strategy approaches ought to be encouraged to address one or more concomitant demands such as fat and salt reduction with efficient sensory, technological, and microbiological quality. It also emphasizes the feasibility of combined use of alternative ingredients and emerging technologies aiming at microbiological safety to healthier products.

REFERENCES

[1]   L.C. Shan, A. De Brún, M. Henchion, C. Li, C. Murrin, P.G. Wall, and F.J. Monahan, "Consumer evaluations of processed meat products reformulated to be healthier - A conjoint analysis study", *Meat Sci.,* vol. 131, pp. 82-89, 2017.
[http://dx.doi.org/10.1016/j.meatsci.2017.04.239] [PMID: 28494317]

[2]   Z.L. Kang, F.S. Chen, and H.J. Ma, "Effect of pre-emulsified soy oil with soy protein isolate in frankfurters: A physical-chemical and Raman spectroscopy study", *Lebensm. Wiss. Technol.,* vol. 74, pp. 465-471, 2016.
[http://dx.doi.org/10.1016/j.lwt.2016.08.011]

[3]   D.T. Utama, H.S. Jeong, J. Kim, F.H. Barido, and S.K. Lee, "Fatty acid composition and quality properties of chicken sausage formulated with pre-emulsified perilla-canola oil as an animal fat replacer", *Poult. Sci.,* vol. 98, no. 7, pp. 3059-3066, 2019.
[http://dx.doi.org/10.3382/ps/pez105] [PMID: 30877751]

[4]   H.Y. Hu, J. Pereira, L.J. Xing, Y.Y. Hu, C.L. Qiao, G.H. Zhou, and W.G. Zhang, "Effects of regenerated cellulose emulsion on the quality of emulsified sausage", *LWT - Food Sci. Technol,* vol. 70, pp. 315-321, 2016.
[http://dx.doi.org/10.1016/j.lwt.2016.02.055]

[5] T. Pintado, A.M. Herrero, F. Jiménez-Colmenero, and C. Ruiz-Capillas, "Strategies for incorporation of chia (Salvia hispanica L.) in frankfurters as a health-promoting ingredient", *Meat Sci.,* vol. 114, pp. 75-84, 2016.
[http://dx.doi.org/10.1016/j.meatsci.2015.12.009] [PMID: 26745305]

[6] T.L. Wolfer, N.C. Acevedo, K.J. Prusa, J.G. Sebranek, and R. Tarté, "Replacement of pork fat in frankfurter-type sausages by soybean oil oleogels structured with rice bran wax", *Meat Sci.,* vol. 145, pp. 352-362, 2018.
[http://dx.doi.org/10.1016/j.meatsci.2018.07.012] [PMID: 30031201]

[7] M. Alejandre, D. Passarini, I. Astiasarán, and D. Ansorena, "The effect of low-fat beef patties formulated with a low-energy fat analogue enriched in long-chain polyunsaturated fatty acids on lipid oxidation and sensory attributes", *Meat Sci.,* vol. 134, pp. 7-13, 2017.
[http://dx.doi.org/10.1016/j.meatsci.2017.07.009]

[8] M. Alejandre, I. Astiasarán, D. Ansorena, and S. Barbut, "Using canola oil hydrogels and organogels to reduce saturated animal fat in meat batters", *Food Res. Int.,* vol. 122, pp. 129-136, 2019.
[http://dx.doi.org/10.1016/j.foodres.2019.03.056] [PMID: 31229064]

[9] I. Oh, J.H. Lee, H.G. Lee, and S. Lee, "Feasibility of hydroxypropyl methylcellulose oleogel as an animal fat replacer for meat patties", *Food Res. Int.,* vol. 122, pp. 566-572, 2019.
[http://dx.doi.org/10.1016/j.foodres.2019.01.012]

[10] D. Kouzounis, A. Lazaridou, and E. Katsanidis, "Partial replacement of animal fat by oleogels structured with monoglycerides and phytosterols in frankfurter sausages", *Meat Sci.,* vol. 130, pp. 38-46, 2017.
[http://dx.doi.org/10.1016/j.meatsci.2017.04.004] [PMID: 28407498]

[11] S. Barbut, J. Wood, and A. Marangoni, "Potential use of organogels to replace animal fat in comminuted meat products", *Meat Sci.,* vol. 122, pp. 155-162, 2016.
[http://dx.doi.org/10.1016/j.meatsci.2016.08.003] [PMID: 27552678]

[12] T. Pintado, A.M. Herrero, F. Jiménez-Colmenero, C. Pasqualin Cavalheiro, and C. Ruiz-Capillas, "Chia and oat emulsion gels as new animal fat replacers and healthy bioactive sources in fresh sausage formulation", *Meat Sci,* vol. 135, pp. 6-13, 2018.
[http://dx.doi.org/10.1016/j.meatsci.2017.08.004]

[13] I. Tavernier, A.R. Patel, P. Van der Meeren, and K. Dewettinck, "Emulsion-templated liquid oil structuring with soy protein and soy protein: κ-carrageenan complexes", *Food Hydrocoll.,* vol. 65, pp. 107-120, 2017.
[http://dx.doi.org/10.1016/j.foodhyd.2016.11.008]

[14] R.T. Heck, R.G. Vendruscolo, M. de Araújo Etchepare, A.J. Cichoski, C.R. de Menezes, J.S. Barin, J.M. Lorenzo, R. Wagner, and P.C.B. Campagnol, "Is it possible to produce a low-fat burger with a healthy n-6/n-3 PUFA ratio without affecting the technological and sensory properties?", *Meat Sci.,* vol. 130, pp. 16-25, 2017.
[http://dx.doi.org/10.1016/j.meatsci.2017.03.010] [PMID: 28347883]

[15] L. Salcedo-Sandoval, S. Cofrades, C. Ruiz-Capillas, A. Matalanis, D.J. McClements, E.A. Decker, and F. Jiménez-Colmenero, "Oxidative stability of n-3 fatty acids encapsulated in filled hydrogel particles and of pork meat systems containing them", *Food Chem.,* vol. 184, pp. 207-213, 2015.
[http://dx.doi.org/10.1016/j.foodchem.2015.03.093] [PMID: 25872446]

[16] L. Salcedo-Sandoval, S. Cofrades, C. Ruiz-Capillas, and F. Jiménez-Colmenero, "Filled hydrogel particles as a delivery system for n-3 long chain PUFA in low-fat frankfurters: Consequences for product characteristics with special reference to lipid oxidation", *Meat Sci.,* vol. 110, pp. 160-168, 2015.
[http://dx.doi.org/10.1016/j.meatsci.2015.07.013] [PMID: 26232749]

[17] J.H. Choe, H.Y. Kim, J.M. Lee, Y.J. Kim, and C.J. Kim, "Quality of frankfurter-type sausages with added pig skin and wheat fiber mixture as fat replacers", *Meat Sci.,* vol. 93, no. 4, pp. 849-854, 2013.

[http://dx.doi.org/10.1016/j.meatsci.2012.11.054] [PMID: 23313971]

[18]   L.A.A.S. Alves, J. M. Lorenzo, C.A.A. Gonçalves, B.A. Santos, R.T. Heck, A.J. Cichoski, and P.C.B. Campagnol, "Production of healthier bologna type sausages using pork skin and green banana flour as a fat replacers", *Meat Sci,* vol. 121, pp. 73-78, 2016.
[http://dx.doi.org/10.1016/j.meatsci.2016.06.001]

[19]   Y.S. Choi, H.W. Kim, K.E. Hwang, D.H. Song, J.H. Choi, M.A. Lee, H.J. Chung, and C.J. Kim, "Physicochemical properties and sensory characteristics of reduced-fat frankfurters with pork back fat replaced by dietary fiber extracted from makgeolli lees", *Meat Sci.,* vol. 96, pp. 892-900, 2014.
[http://dx.doi.org/10.1016/j.meatsci.2013.08.033] [PMID: 24200582]

[20]   M. Han, M.P. Clausen, M. Christensen, E. Vossen, T. Van Hecke, and H.C. Bertram, "Enhancing the health potential of processed meat: the effect of chitosan or carboxymethyl cellulose enrichment on inherent microstructure, water mobility and oxidation in a meat-based food matrix", *Food Funct.,* vol. 9, no. 7, pp. 4017-4027, 2018.
[http://dx.doi.org/10.1039/C8FO00835C] [PMID: 29978871]

[21]   H.W. Kim, D. Setyabrata, Y.J. Lee, and Y.H. Brad Kim, "Efficacy of alkali-treated sugarcane fiber for improving physicochemical and textural properties of meat emulsions with different fat levels", *Han-gug Chugsan Sigpum Hag-hoeji,* vol. 38, no. 2, pp. 315-324, 2018.
[PMID: 29805281]

[22]   N.J. Oswell, H. Thippareddi, and R.B. Pegg, "Practical use of natural antioxidants in meat products in the U.S.: A review", *Meat Sci.,* vol. 145, pp. 469-479, 2018.
[http://dx.doi.org/10.1016/j.meatsci.2018.07.020] [PMID: 30071458]

[23]   Y. Kumar, D.N. Yadav, T. Ahmad, and K. Narsaiah, "Recent trends in the use of natural antioxidants for meat and meat products", *Compr. Rev. Food Sci. Food Saf.,* vol. 14, no. 6, pp. 796-812, 2015.
[http://dx.doi.org/10.1111/1541-4337.12156]

[24]   M. Pateiro, F.J. Barba, R. Domínguez, A.S. Sant'Ana, A. Mousavi Khaneghah, M. Gavahian, B. Gómez, and J.M. Lorenzo, "Essential oils as natural additives to prevent oxidation reactions in meat and meat products: A review", *Food Res. Int.,* vol. 113, pp. 156-166, 2018.
[http://dx.doi.org/10.1016/j.foodres.2018.07.014] [PMID: 30195508]

[25]   M. Aminzare, "Using natural anti-oxidants in meat and meat products as preservatives: A review", *Adv. Anim. Vet. Sci.,* vol. 7, no. 5, pp. 417-426, 2019.
[http://dx.doi.org/10.17582/journal.aavs/2019/7.5.417.426]

[26]   V. Tomović, M. Jokanović, B. Šojić, S. Škaljac, and M. Ivić, "Plants as natural anti-oxidants for meat products", *IOP Conf. Ser. Earth Environ. Sci.,* vol. 85, no. 1, 2017.
[http://dx.doi.org/10.1088/1755-1315/85/1/012030]

[27]   A.I. Andrés, M.J. Petrón, J.D. Adámez, M. López, and M.L. Timón, "Food by-products as potential antioxidant and antimicrobial additives in chill stored raw lamb patties", *Meat Sci.,* vol. 129, pp. 62-70, 2017.
[http://dx.doi.org/10.1016/j.meatsci.2017.02.013] [PMID: 28259073]

[28]   H.S. Kim, and K.B. Chin, "Evaluation of antioxidative activity of various levels of ethanol extracted tomato powder and application to pork patties", *Han-gug Chugsan Sigpum Hag-hoeji,* vol. 37, no. 2, pp. 242-253, 2017.
[http://dx.doi.org/10.5851/kosfa.2017.37.2.242] [PMID: 28515648]

[29]   F.B. Brum, F.T. Macagnan, M.A. Monego, T. André, and L. Picolli, "Phytic acid addition into hamburger-type meat product", *Rev. Inst. Adolfo Lutz,* vol. 70, no. 1, pp. 47-52, 2011.

[30]   J. Choe, Y.H.B. Kim, H.Y. Kim, and C.J. Kim, "Evaluations of physicochemical and anti-oxidant properties of powdered leaves from lotus, shepherd's purse and goldenrod in restructured duck/pork patties", *J. Food Sci. Technol.,* vol. 54, no. 8, pp. 2494-2502, 2017.
[http://dx.doi.org/10.1007/s13197-017-2693-6] [PMID: 28740307]

[31]   S. Ahmad Mir, F. Ahmad Masoodi, and J. Raja, "Influence of natural anti-oxidants on microbial load, lipid oxidation and sensorial quality of rista—A traditional meat product of India", *Food Biosci.,* vol. 20, pp. 79-87, 2017.
[http://dx.doi.org/10.1016/j.fbio.2017.08.004]

[32]   H.M. Dilnawaz, S. Kumar, and Z.F. Bhat, "Ipomoea batatas as a novel binding agent for hot-set restructured binding systems and green coffee bean for improved lipid oxidative stability and storage quality", *Nutr. Food Sci.,* vol. 47, no. 5, pp. 659-672, 2017.
[http://dx.doi.org/10.1108/NFS-04-2017-0066]

[33]   J.H. Kim, D.U. Ahn, J.B. Eun, and S.H. Moon, "Anti-oxidant effect of extracts from the coffee residue in raw and cooked meat", *Antioxidants,* vol. 5, no. 3, pp. 1-5, 2016.
[http://dx.doi.org/10.3390/antiox5030021] [PMID: 27384587]

[34]   P.E.S. Munekata, R. Domínguez, D. Franco, R. Bermúdez, M.A. Trindade, and J.M. Lorenzo, "Effect of natural antioxidants in Spanish salchichón elaborated with encapsulated n-3 long chain fatty acids in konjac glucomannan matrix", *Meat Sci.,* vol. 124, pp. 54-60, 2017.
[http://dx.doi.org/10.1016/j.meatsci.2016.11.002] [PMID: 27835835]

[35]   S. Zamuz, M. López-Pedrouso, F.J. Barba, J.M. Lorenzo, H. Domínguez, and D. Franco, "Application of hull, bur and leaf chestnut extracts on the shelf-life of beef patties stored under MAP: Evaluation of their impact on physicochemical properties, lipid oxidation, antioxidant, and antimicrobial potential", *Food Res. Int.,* vol. 112, pp. 263-273, 2018.
[http://dx.doi.org/10.1016/j.foodres.2018.06.053] [PMID: 30131137]

[36]   J.C. Baldin, P.E.S. Munekata, E.C. Michelin, Y.J. Polizer, P.M. Silva, T.M. Canan, M.A. Pires, S.H.S. Godoy, C.S. Fávaro-Trindade, C.G. Lima, A.M. Fernandes, and M.A. Trindade, "Effect of microencapsulated Jabuticaba (Myrciaria cauliflora) extract on quality and storage stability of mortadella sausage", *Food Res. Int.,* vol. 108, pp. 551-557, 2018.
[http://dx.doi.org/10.1016/j.foodres.2018.03.076] [PMID: 29735090]

[37]   J.M. Lorenzo, F.C. Vargas, I. Strozzi, M. Pateiro, M.M. Furtado, A.S. Sant'Ana, G. Rocchetti, F.J. Barba, R. Dominguez, L. Lucini, and P.J. do Amaral Sobral, "Influence of pitanga leaf extracts on lipid and protein oxidation of pork burger during shelf-life", *Food Res. Int.,* vol. 114, pp. 47-54, 2018.
[http://dx.doi.org/10.1016/j.foodres.2018.07.046] [PMID: 30361026]

[38]   M. Pateiro, F.C. Vargas, A.A.I.A. Chincha, A.S. Sant'Ana, I. Strozzi, G. Rocchetti, F.J. Barba, R. Domínguez, L. Lucini, P.J. do Amaral Sobral, and J.M. Lorenzo, "Guarana seed extracts as a useful strategy to extend the shelf life of pork patties: UHPLC-ESI/QTOF phenolic profile and impact on microbial inactivation, lipid and protein oxidation and antioxidant capacity", *Food Res. Int.,* vol. 114, pp. 55-63, 2018.
[http://dx.doi.org/10.1016/j.foodres.2018.07.047] [PMID: 30361027]

[39]   J.S. Ribeiro, M.J.M.C Santos, L.K.R. Silva, L.C.L. Pereira, I.A. Santos, S.C.S. Lannes, and M.V. Silva, "Natural anti-oxidants used in meat products: A brief review", *Meat Sci,* vol. 148, pp. 181-188, 2019.
[http://dx.doi.org/10.1016/j.meatsci.2018.10.016]

[40]   E. Desmond, "Reducing salt: A challenge for the meat industry", *Meat Sci.,* vol. 74, no. 1, pp. 188-196, 2006.
[http://dx.doi.org/10.1016/j.meatsci.2006.04.014] [PMID: 22062728]

[41]   A. Grossi, J. Søltoft-Jensen, J.C. Knudsen, M. Christensen, and V. Orlien, "Reduction of salt in pork sausages by the addition of carrot fibre or potato starch and high pressure treatment", *Meat Sci.,* vol. 92, no. 4, pp. 481-489, 2012.
[http://dx.doi.org/10.1016/j.meatsci.2012.05.015] [PMID: 22682686]

[42]   E.S. Inguglia, Z. Zhang, B.K. Tiwari, J.P. Kerry, and C.M. Burgess, "Salt reduction strategies in processed meat products – A review", *Trends Food Sci. Technol.,* vol. 59, pp. 70-78, 2017.
[http://dx.doi.org/10.1016/j.tifs.2016.10.016]

[43]   M. Armenteros, M.C. Aristoy, J.M. Barat, and F. Toldrá, "Biochemical and sensory changes in dry-cured ham salted with partial replacements of NaCl by other chloride salts", *Meat Sci.,* vol. 90, no. 2, pp. 361-367, 2012.
[http://dx.doi.org/10.1016/j.meatsci.2011.07.023] [PMID: 21871742]

[44]   V.A.S. Vidal, J.P. Biachi, C.S. Paglarini, M.B. Pinton, P.C.B. Campagnol, E.A. Esmerino, A.G. da Cruz, M.A. Morgano, and M.A.R. Pollonio, "Reducing 50% sodium chloride in healthier jerked beef: An efficient design to ensure suitable stability, technological and sensory properties", *Meat Sci.,* vol. 152, pp. 49-57, 2019.
[http://dx.doi.org/10.1016/j.meatsci.2019.02.005] [PMID: 30802818]

[45]   M.A. Pires, P.E. S. Munekata, J.C. Baldin, Y.J.P. Rocha, L.T. Carvalho, I.R. dos Santos, J.C. Barros, M.A. Trindade, "The effect of sodium reduction on the microstructure, texture and sensory acceptance of Bologna sausage", *Food Struct.,* vol. 14, pp. 1-7, 2017.
[http://dx.doi.org/10.1016/j.foostr.2017.05.002]

[46]   M. Triki, I. Khemakhem, I. Trigui, R. Ben Salah, S. Jaballi, C. Ruiz-Capillas, M.A. Ayadi, H. Attia, and S. Besbes, "Free-sodium salts mixture and AlgySalt® use as NaCl substitutes in fresh and cooked meat products intended for the hypertensive population", *Meat Sci.,* vol. 133, pp. 194-203, 2017.
[http://dx.doi.org/10.1016/j.meatsci.2017.07.005] [PMID: 28715706]

[47]   D. Seganfredo, S. Rodrigues, D.L. Kalschne, C.M.P. Sarmento, and C. Canan, "Partial substitution of sodium chloride in Toscana sausages and the effect on product characteristics", *Semin. Agrar.,* vol. 37, no. 3, pp. 1285-1294, 2016.
[http://dx.doi.org/10.5433/1679-0359.2016v37n3p1285]

[48]   G.O. Cox, and S.Y. Lee, "Sodium threshold in model reduced and low fat oil-in-water emulsion systems", *J. Food Sci.,* vol. 83, no. 3, pp. 791-797, 2018.
[http://dx.doi.org/10.1111/1750-3841.13941] [PMID: 29509976]

[49]   E. Atilgan, and B. Kilic, "Effects of microbial transglutaminase, fibrimex and alginate on physicochemical properties of cooked ground meat with reduced salt level", *J. Food Sci. Technol.,* vol. 54, no. 2, pp. 303-312, 2017.
[http://dx.doi.org/10.1007/s13197-016-2463-x] [PMID: 28242929]

[50]   M. Viuda-Martos, J. Fernández-López, E. Sayas-Barbera, E. Sendra, C. Navarro, and J.A. Pérez-Alvarez, "Citrus co-products as technological strategy to reduce residual nitrite content in meat products", *J. Food Sci.,* vol. 74, no. 8, pp. R93-R100, 2009.
[http://dx.doi.org/10.1111/j.1750-3841.2009.01334.x] [PMID: 19799678]

[51]   J.G. Sebranek, "Basic curing ingredients", In: *Ingredients in Meat Products: Properties, Functionality and Applications.,* R. Tarté, Ed., Springer Science, 2010, pp. 1-24.

[52]   E. De Mey, H. De Maere, H. Paelinck, and I. Fraeye, "Volatile N-nitrosamines in meat products: Potential precursors, influence of processing, and mitigation strategies", *Crit. Rev. Food Sci. Nutr.,* vol. 57, no. 13, pp. 2909-2923, 2017.
[http://dx.doi.org/10.1080/10408398.2015.1078769] [PMID: 26528731]

[53]   J.J. Sindelar, J.C. Cordray, J.G. Sebranek, J.A. Love, and D.U. Ahn, "Effects of varying levels of vegetable juice powder and incubation time on color, residual nitrate and nitrite, pigment, pH, and trained sensory attributes of ready-to-eat uncured ham", *J. Food Sci.,* vol. 72, no. 6, pp. S388-S395, 2007.
[http://dx.doi.org/10.1111/j.1750-3841.2007.00404.x] [PMID: 17995695]

[54]   S. Tahmouzi, S.H. Razavi, M. Safari, and Z. Emam-Djomeh, "Development of a practical method for processing of nitrite-free hot dogs with emphasis on evaluation of physico-chemical and microbiological properties of the final product during refrigeration", *J. Food Process. Preserv.,* vol. 37, no. 2, pp. 109-119, 2013.
[http://dx.doi.org/10.1111/j.1745-4549.2011.00626.x]

[55]   C. Ruiz-Capillas, S. Tahmouzi, M. Triki, L. Rodríguez-Salas, F. Jiménez-Colmenero, and A.M.

Herrero, "Nitrite-free Asian hot dog sausages reformulated with nitrite replacers", *J. Food Sci. Technol.,* vol. 52, no. 7, pp. 4333-4341, 2015.
[http://dx.doi.org/10.1007/s13197-014-1460-1] [PMID: 26139898]

[56]   C. Ruiz-Capillas, A.M. Herrero, S. Tahmouzi, S.H. Razavi, M. Triki, L. Rodríguez-Salas, K. Samcová, and F. Jiménez-Colmenero, "Properties of reformulated hot dog sausage without added nitrites during chilled storage", *Food Sci. Technol. Int.,* vol. 22, no. 1, pp. 21-30, 2016.
[http://dx.doi.org/10.1177/1082013214562919] [PMID: 25480689]

[57]   C. Aquilani, F. Sirtori, M. Flores, R. Bozzi, B. Lebret, and C. Pugliese, "Effect of natural antioxidants from grape seed and chestnut in combination with hydroxytyrosol, as sodium nitrite substitutes in Cinta Senese dry-fermented sausages", *Meat Sci.,* vol. 145, pp. 389-398, 2018.
[http://dx.doi.org/10.1016/j.meatsci.2018.07.019] [PMID: 30036844]

[58]   M. García, T. Beldarraín, L. Fornaris, and R. Díaz, "Partial substitution of nitrite by chitosan and the effect on the quality properties of pork sausages", *Food Sci. Technol. (Campinas),* vol. 31, no. 2, pp. 481-487, 2011.
[http://dx.doi.org/10.1590/S0101-20612011000200031]

[59]   C. Canan, A.P.M. Bloot, I. Baraldi, E. Colla, V.P. Feltrin, M.P. Corso, S. Habu, D.R.N. Nogues, D.L. Kalschne, and R.A. Silva-Buzanello, "Composição antimicrobiana a base de fitato, e seu uso", *BR1020180062280,*   2018.   https://gru.inpi.gov.br/pePI/   servlet/ PatenteServletController?Action=detail&CodPedido=1471693&SearchParameter=BR1020180062280 %20%20%20%20%20%20&Resumo=&Titulo=

SUBJECT INDEX

A

Acid 10, 12, 16, 29, 35, 60, 67, 79, 105, 117, 118
  acetic 67
  ascorbic 10, 117, 118
  ferulic 118
  polyricinoleic 35
  phytic 117
  sodium 29
  thiobarbituric 16, 79
Acoustic 31, 34, 68, 85, 86, 93
  cavitation 31, 34
  energy 85
  parameters 86, 93
  waves 68
Action 11, 88, 120, 121
  conservative 120
  efficient antimicrobial 121
  lysosomal enzymes 88
  proteolytic 88
  starter cultures 11
Additives 10, 27, 59, 60, 62, 64, 76, 113, 114, 118, 119, 121
  chemical 76, 113, 114
  natural 118
  prospective 119
  starter culture reference 10
Alignment 105, 107, 108
  biopolymer solutions 105
  freeze 107
  ice crystal needles 108
Alkaline phosphate mixtures 60
Alterations 38, 80, 93, 95
  structural 95
Analogues production 103, 107
  whole-muscle meat 107
Animal fat 5, 27, 32, 33, 34, 35, 36, 115
  replacing 32, 33, 34, 35
Animal meat 101, 102, 105, 107, 109
  simulate 101
  structures 105
Anisotropic protein 105
  producing 105

Anti-oxidants 117
  effective 117
  natural herbal 117
Aqueous 26, 94
  salt 26
  solutions 94
Aroma 1, 10, 11, 17, 29, 63, 87, 109, 115
  characteristic 63
  commercial 29
  natural smoked 29
Attributes 19, 20, 38, 49, 69, 85, 87, 102
  key 85
  negative 69
  sensory 38, 49, 102

B

Bacillus subtilis 14
Bacon 4, 45, 48, 53
  cured 48
  producing 53
Bacteria 15, 19, 22, 34, 52, 67, 69, 87, 96
  lactic 34
  pathogenic 69
  psychrophilic 96
  psychrotrophic 67
Bacteria inhibition 18
B-complex vitamins 4
Beef 32, 33, 68, 115, 116
  fat 115, 116
  frozen 68
  myofibrillar proteins 32, 33
  separated chicken 29
  unfrozen 68
Beef tenderness 85, 86, 87, 88, 96
  effects of ultrasound on 85
Behavior 3, 51, 80, 120
  influence consumer 3
  textural 120
Biological 13, 101
  materials 13
Biopolymer blends 102, 105
  structuring 102
Bologna sausage 27, 28, 29, 32, 34, 39

Daneysa L. Kalschne, Marinês P. Corso & Cristiane Canan

www.ingramcontent.com/pod-product-compliance
Lightning Source LLC
Chambersburg PA
CBHW041714210326
41598CB00007B/644